S0-DSZ-609

Published 1982 by Warwick Press
730 Fifth Avenue, New York, New York 10019.

First published in Great Britain by
Longman Group Limited 1982.

Printed in Italy by Vallardi Industrie Grafiche.

6 5 4 3 2 1

Library of Congress Catalog Card No. 81-51794

ISBN 0-531-09196-1

UNDERSTANDING SCIENCE

Lasers

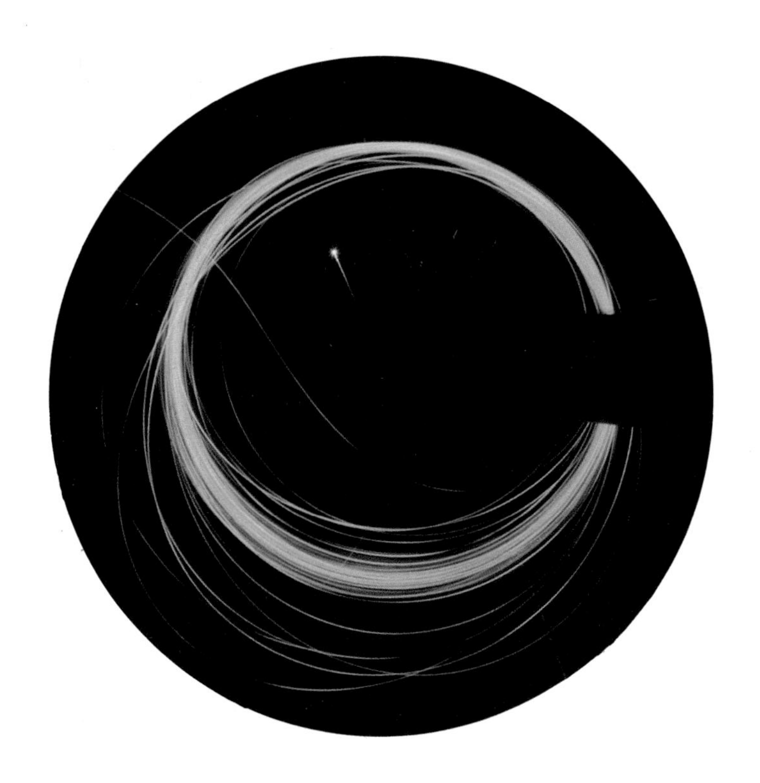

WARWICK PRESS

CONTENTS

Author
William Burroughs

Editors
Mark Lambert
John Paton

FOREWORD

Thirty years ago lasers were just a futuristic dream. Today they are very real and are becoming an increasingly important part of our lives. Developments in laser technology are taking place very rapidly. In industry and research centres in many parts of the world new and better lasers are being devised at an amazing rate.

Lasers, with their very special properties, have revolutionized many areas of science and technology. Measurements of such things as distance and speed can now be made more accurately than ever seemed possible. Using lasers, scientists have vastly improved a number of chemical techniques and have begun to learn the details of how chemical reactions work. High power lasers provide fast, accurate and clean cutting tools for use in industry and medicine.

The use of laser light, together with optical fibres, is becoming increasingly important in the communications industry. Within the next few decades perhaps all telephone conversations and business information will be transmitted by lasers. Finally, laser light can be used to produce the most amazing three-dimensional images. One day, this may lead to three-dimensional television and video telephone pictures.

In this book you will find out how lasers work and how their unique properties are used in all these applications. However, this is just the beginning. Laser technology has an exciting future and many new and wonderful developments will occur in the next few years.

The Age of Lasers

The laser is one of the most dramatic developments of the 20th century. It is a device that emits a narrow beam of light. But this is no ordinary light; it has very special properties, and lasers can be used to perform a wide range of different tasks. They can be used to cut metals at high speed and to carry out delicate machining and drilling work on hard and brittle materials. Finely controlled lasers have been used to perform delicate surgery and dentistry. They have been used to measure the speed of light with incredible accuracy, and to detect tiny amounts of distant pollutants in the atmosphere. The list of uses is increasing all the time.

All these and many other developments have taken place in little more than 20 years. The first laser was built in 1960 by Theodore Maiman and other scientists at the research laboratories of the Hughes Aircraft Company in California. It contained a special man-made ruby rod, around which was coiled a flash tube. When the ruby rod was subjected to intense flashes of ordinary white light, it produced pulses of red laser light.

This breakthrough did not come as a complete surprise. The basic theory of the laser had been put forward by Albert Einstein in 1917. What was missing was the equipment to build lasers. Only with the rapid development of technology after World War II could the necessary equipment be built.

Once the initial discovery had been made, the flood-gates opened. Within a few years a whole host of different lasers, had been produced, using solids, liquids and gases. The brilliant colours of laser light, combined with claims for its amazing properties, attracted widespread attention.

The possibility of a laser "death ray", already popular in science fiction, was immediately taken up in films. In the film *Goldfinger* James Bond was threatened with a terrible end as an "industrial" laser sliced its way through a slab of gold towards him. This image of the laser has grown and lasers have come to be regarded as the weapons of the future. The visual appeal of laser light was also exploited; pop groups and open air art shows were soon using brilliantly coloured laser displays.

The Laser Bears Fruit

In fact, laser developments did not quite live up to all this publicity. Scientists began to discover more and more uses for lasers. But for the first ten years or so the pace of discovery was so fast that the technologists could not keep up. Then, during the 1970s some of the early discoveries began to bear fruit.

This progress has been marked by many spectacular successes. The power of an early laser was used to project a visible spot of light onto the Moon. When the Apollo II astronauts landed on the Moon in 1969, they left behind a special mirror which has since been used to measure the distance between Earth and the Moon to within a few centimetres. Immediate use was made of high power lasers in medicine and dentistry. And as power levels increased, heavier tasks

Lasers are used extensively in drilling operations where great accuracy is required. Here a hole is being drilled in aluminium.

were taken on in cutting metals. In the last few years the possibility of a real "death ray" has begun to emerge as intensive work in the USA and the Soviet Union has raised power levels.

The more gentle measuring properties of laser light have also been exploited. The properties of atoms and molecules can now be measured with an accuracy undreamed of before the advent of lasers. The speed of light has been measured with a tremendous increase in accuracy. The combination of these measuring properties with high power levels now offers the hope of a totally new form of chemistry.

Less complicated uses for lasers have been found in construction work. A pencil-thin laser beam can be used to align parts of bridges and huge buildings with great precision. Even simple tasks, such as digging trenches and laying pipelines are made easier. Lasers are also becoming an important part of new telecommunication systems. They are being used not only to transmit data but also to read it and write it.

All these developments make use of the special properties of laser light and the wide variety of materials that can be used in lasers. The fact that a large number of gases, liquids and solids can be made to produce laser light is important. Each different material produces a different form of light. So the best laser for any given task can be found from the wide range available. Lasers also come in many sizes. Huge multi-million pound systems are being built to see if they offer a way of producing limitless power by atomic fusion. At the other extreme, solid lasers no larger than a grain of salt have been made for telephone systems.

The age of the laser is now upon us.

The laser first made its impact on the public because of its brilliant 'electronic fireworks' displays such as that shown here in a concert of The Who.

A New Form of Light

Light is a familiar part of everyday life. It is a form of energy and like all forms of energy it has its own special properties. Laser light has particularly special properties, but to understand these it is first necessary to understand some of the properties of ordinary light.

Light as a Wave

A beam of light can be described as consisting of a series of waves. like the waves on an ocean. It has a *wavelength,* which is simply the distance between the crests or troughs of successive waves. It has a *frequency*, which is the number of waves that pass a given point in one second (this is also described as the number of cycles per second). And the waves travel at a constant speed. Because this is so, the distance a wave travels in one second is the wavelength multiplied by the number of waves that pass. This relationship can be expressed in the formula $c = \lambda \times f$, in which c is the speed of light measured in metres per second, λ (the Greek letter lambda) is the wavelength measured in metres and f is the frequency measured in cycles per second, or hertz. Notice that, because the speed of light (c) is constant, as the wavelength (λ) increases the frequency (f) must decrease. So high frequencies go with short wavelengths and vice versa.

But there are important differences between ocean waves and light waves. The waves of an ocean are many metres apart and they travel at only a few metres per second. Light, on the other hand, has a wavelength of only about half a micrometre (one millionth of a metre). It has a very high frequency—some 500 million cycles per second. And, most importantly, it travels at an incredible speed—nearly 300,000 kilometres per second.

Light waves are described as electromagnetic waves, because they have both electrical and magnetic components. But light is not the only kind of electromagnetic wave. In fact, the light our eyes can see makes up only a small part of a whole range of different wavelengths. The range is known as the electromagnetic spectrum, which stretches from long wavelength radio waves, through microwaves, infra-red rays, light rays, ultra-violet rays and X-rays to very short wavelength

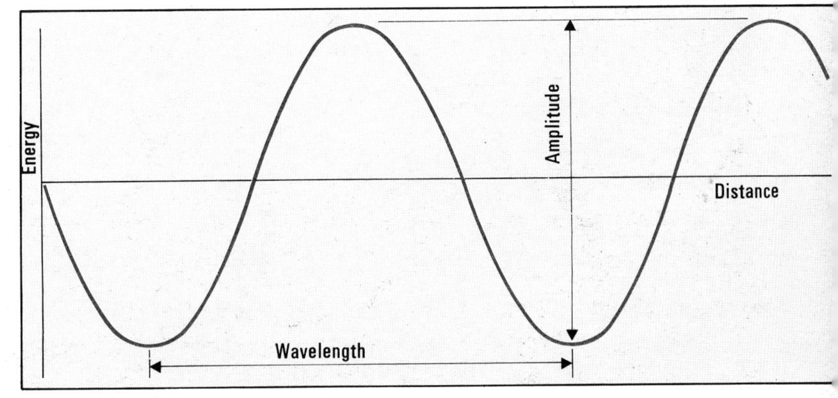

Light is visualized as a series of waves which travel at the unimaginably high speed of 300,000 kilometres per second. This speed is constant. As the frequency of the waves passing a given point rises, the distance between wave peaks decreases.

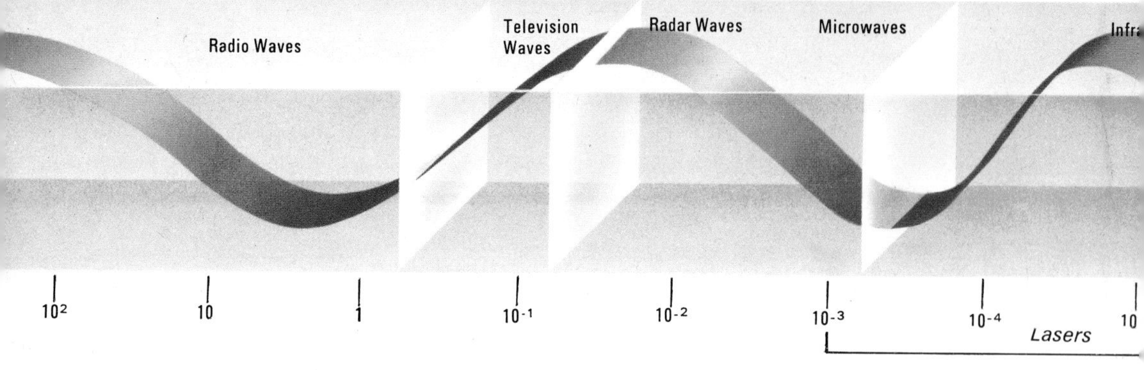

gamma rays and cosmic rays. This is important to remember because not all lasers produce light; they can be made to work over a large part of the electromagnetic spectrum, from wavelengths of about a millimetre to wavelengths of less than one ten-thousandth of a millimetre—most of which is not visible to the human eye.

One other property of light is important in lasers. This is the intensity, or brightness, of light, which is related to the amplitude of the wave (the total height of the wave from the top of a crest to the bottom of a trough). The greater the amplitude, the more intense the beam of light.

Light from lasers can be described in the same way as ordinary light. It too has a wavelength and a frequency. But laser light has additional properties. And these are related to the way in which atoms and molecules give out radiation.

The electromagnetic spectrum includes a whole range of electromagnetic waves, from long wavelength (low frequency) radio waves to short wavelength (high frequency) cosmic rays. Visible light forms only a small part of this wide spectrum. And, as yet, lasers can be made to work over only a relatively small range of wavelengths.

Light from Atoms and Molecules

All atoms and molecules (groups of atoms) give off radiation when they are excited by heat, light, electricity of some other form of energy. The way in which a particular atom or molecule does this depends on its structure. Each type of atom or molecule has a different structure and gives out light at only certain wavelengths when it is excited. Therefore, the wavelength of the radiation given off by an excited atom or molecule is a precise measure of its structure. And this useful property enables scientists to study the structure and behaviour of atoms and molecules (see pages 33-37).

The brilliant orange-yellow light produced by a sodium street lamp is a typical example of light given out by excited atoms. The lamp bulb contains sodium atoms. Normally, these atoms are at rest in a "ground state" of energy. But when they are excited by an electrical discharge, their energy state is boosted to a higher level in a series of steps. Then, within a tiny fraction of a second, the atoms tend to fall back towards the ground state. But as they fall from one energy level to another, they release energy in the form of light.

The wavelength of the light given off by an atom depends on the difference between the upper and lower energy levels. The greater the energy difference, the shorter is the wavelength of the radiation given out. In atoms, these energy levels are often quite a long way apart and the light given off therefore has a high frequency and short wavelength. This means that lamps or lasers that work using atoms tend to give off radiation that falls in the near-infra-red, visible or ultra-violet parts of the electromagnetic spectrum—hence the visible orange-yellow light given off by the sodium lamp.

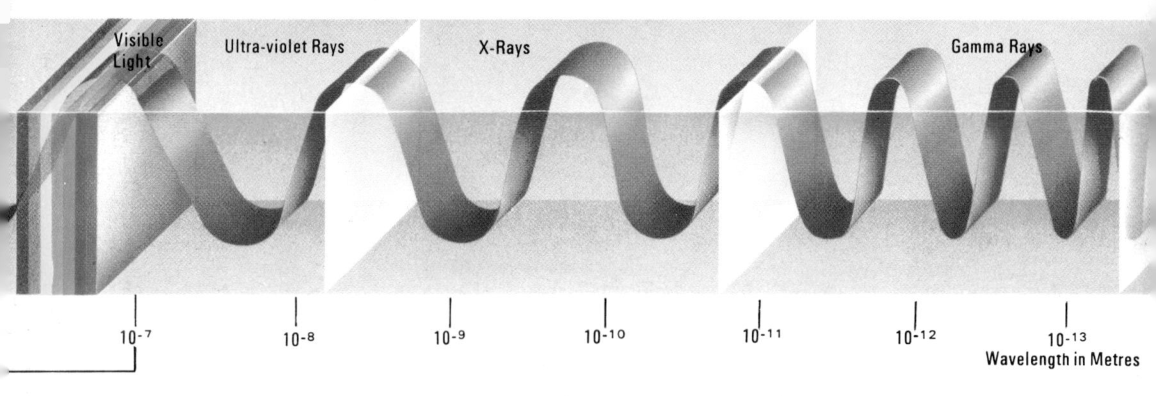

An atom consists of a nucleus, containing protons and neutrons. Round this nucleus travel electrons. When an atom is excited, it is the electrons that receive the energy. Each electron normally orbits the nucleus at a particular level. But when excited, it is boosted to a higher level. Then it falls back to its preferred orbit, releasing energy as it does so.

The electrons of molecules can also be excited to give off high frequency radiation such as light. But molecules also have other energy levels. These are related to the way in which the atoms that make up molecules move in relation to each other. They may move to and from each other (vibrational motion) or around each other (rotational motion). Less energy is involved in vibrational motion than in the excitation of electrons, and even less energy is involved in rotational motion. So radiation connected with these energy levels is usually in the infra-red part of the electromagnetic spectrum.

Laser Light is Special

When an atom or molecule gives out radiation it does so in a very short pulse. This pulse contains the energy released as the atom or molecule returns to a lower energy level. Since the light from a single atom or

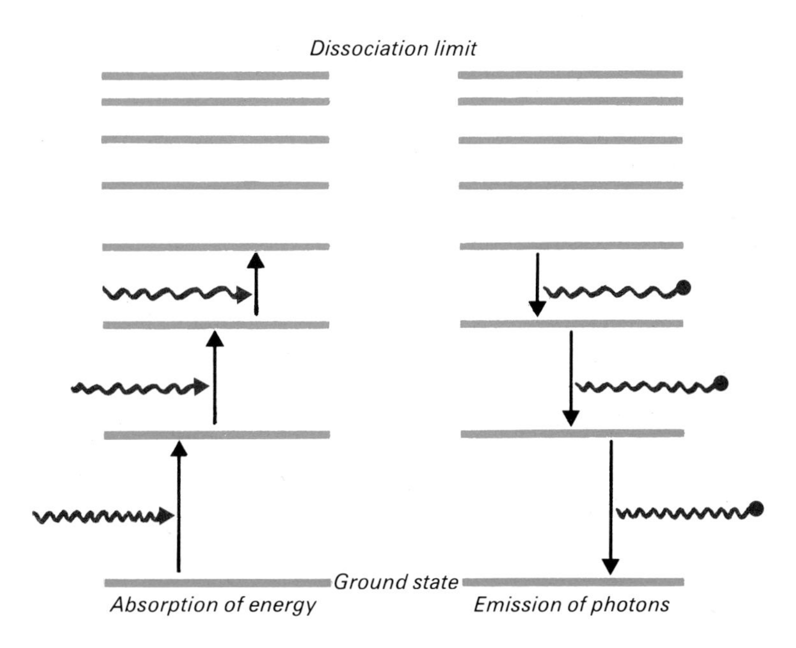

Every type of atom or molecule has its own set of energy levels. At the bottom is the unexcited level, or ground state. The highest level is the point at which the atom or molecule is so excited that it begins to break up, or dissociate. At this dissociation limit an atom loses an electron and a molecule may even lose an atom. Each atom or molecule takes in energy in order to move to a higher energy level (left). Then it loses this energy spontaneously, emitting a photon of electromagnetic radiation, and returns to a lower level (right).

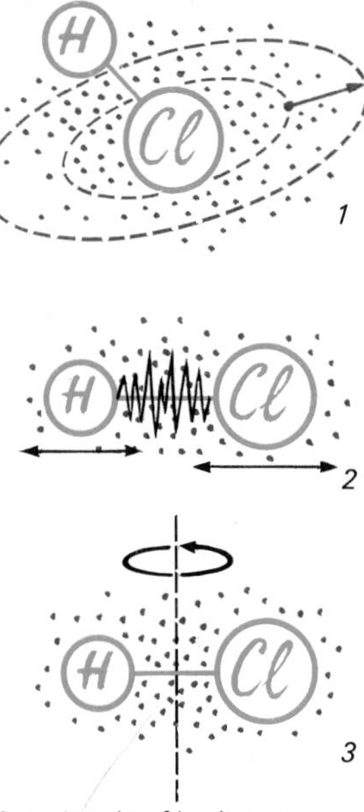

A molecule of hydrogen chloride consists of one hydrogen atom and one chlorine atom. The nuclei of the atoms are surrounded by a cloud of electrons. Such a molecule has three kinds of energy levels. Electronic energy levels **(1)** are those that result from electrons being excited to higher orbits. Vibrational energy levels **(2)** occur as a result of the two atoms vibrating to and from each other. Rotational energy levels **(3)** result from the two atoms rotating around a point between them.

In ordinary light the waves are out of phase and therefore tend to cancel each other out. In laser light, however, the waves are all in phase, thus producing a coherent beam.

molecule has a fixed frequency and wavelength, the pulse can be visualized as a short train of waves and is known as a wave packet, or photon.

In an ordinary lamp each atom fires off its photon of its own accord and does so independently of other atoms. This spontaneous emission means that the individual waves from each atom are not in time with one another. Instead they are all jumbled up. This blurs the wavelength of the light slightly.

In a laser, on the other hand, all the atoms are made to produce photons of identical wavelength and the waves are all in time with each other, or in phase. Light of this kind is called coherent light.

An important difference between the atoms and molecules in an ordinary lamp and those in a laser is the number that occupy the higher energy levels. In an ordinary lamp, when the atoms or molecules are excited, most of them reach the lower energy levels, but only a few reach the highest levels. Of course, each atom or molecule spends only a tiny fraction of a second in an excited state before it emits energy and falls back to a lower level. But at any given moment in a glowing lamp there are more atoms at lower energy levels than at higher levels.

In a laser a different situation occurs. This is because many more atoms or molecules are excited to high energy levels. And once excited to a certain semi-stable level, they become more reluctant to return to the level below and so remain there for slightly longer than at other levels.

When a large number of atoms have been excited to this state, laser action, or lasing, can begin. Take, for example, a situation in which the atoms of the laser material have been excited to a semi-stable level three. Sooner or later one atom starts to fall to level two, releasing a photon as it does so. This photon then strikes a second atom at level three, stimulating it into losing its energy. The second atom also falls to level two, but as it does so it emits another photon of identical wavelength to the first. The two photons move off travelling together in the same direction and exactly in phase. They then strike other atoms and within a fraction of a second the process has been repeated many times. In this way the laser material is made to produce a stream of photons, all of identical wavelengths and all in phase.

Because the waves of light produced by a laser are all in phase, they reinforce, or amplify each other. And it is this process, in which light waves are amplified by stimulating atoms or molecules to emit identical photons, that gives the laser its name. The word "laser" stands for Light Amplification by Stimulated Emission of Radiation.

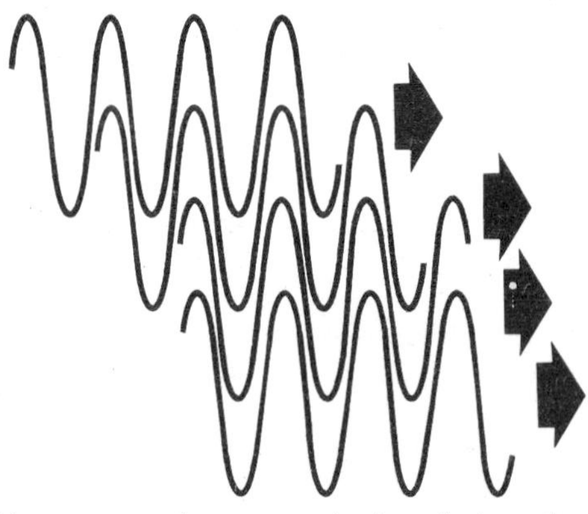
Laser waves in phase (coherent)

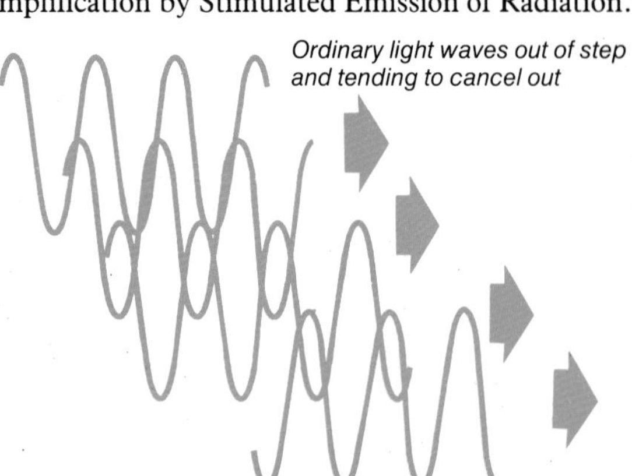
Ordinary light waves out of step and tending to cancel out

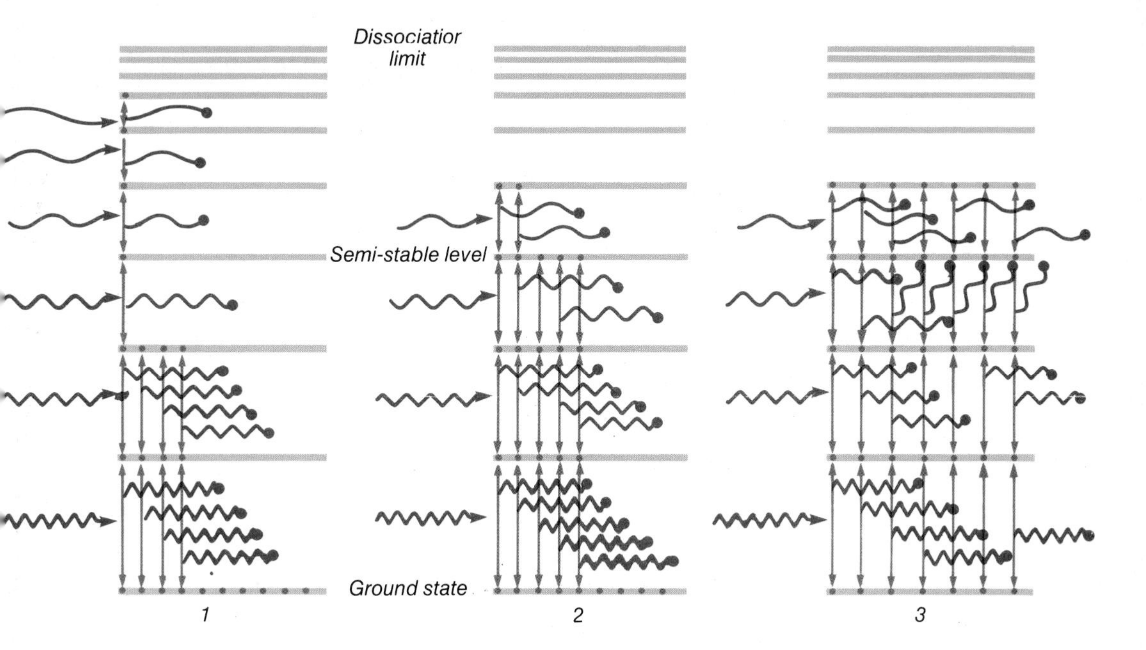

1. When an ordinary lamp is switched on most of the atoms are excited to many energy levels. The excited atoms spontaneously shoot out photons of varying wavelengths. **2.** In a laser, atoms are excited to similar energy states and at a certain semi-stable level they are more reluctant to lose their energy. **3.** Soon, a large number of atoms collect in the semi-stable level. Suddenly, one atom loses its energy. The photon it emits stimulates another atom, and so on. Laser action has begun.

Exciting Laser Materials

The important property of a laser is that it amplifies the light. To produce the greatest amplification of useful light, it is necessary to excite as many atoms or molecules as possible and then ensure that they are all stimulated to emit their energy at the right time. This requires an efficient means of exciting the active material of the laser.

There is a wide variety of materials that can be made to lase. And such materials include solids, liquids and gases. Professor Maiman's first laser used a ruby rod containing a small proportion of chromium atoms, which produced a brilliant red light. High power lasers often use carbon dioxide gas. These different materials have to be excited in different ways.

In gas lasers the most common method of excitation is by using an electrical discharge, like that used in a strip light. Gases can also be excited efficiently by chemical means. This can be done by mixing two flows of different gases which explode when they meet. Such an explosion produces a highly excited new chemical that is well suited for use in very powerful lasers.

Ordinary bright lamps are often used to excite liquids and solids. Sometimes flash tubes, which work in the same way as photographic flash bulbs, are used. Frequently, such lasers give out pulses of light. Some liquid lasers are even excited by other lasers to give different wavelengths. And certain special crystals can emit laser light when a high electric current is passed through them.

Getting Light out of a Laser

Once the active atoms or molecules of a laser material have been excited, it is essential to get as much useful laser light out as possible. This means that when an atom or molecule emits a photon, it must stimulate as many others as possible to send out identical photons. Only in this way can the light, or other radiation, be amplified.

In a solid laser the active material is in the form of a rod. The active materials of gas and liquid lasers are enclosed in rod-shaped containers that resemble strip lights. But there is one major difference between a

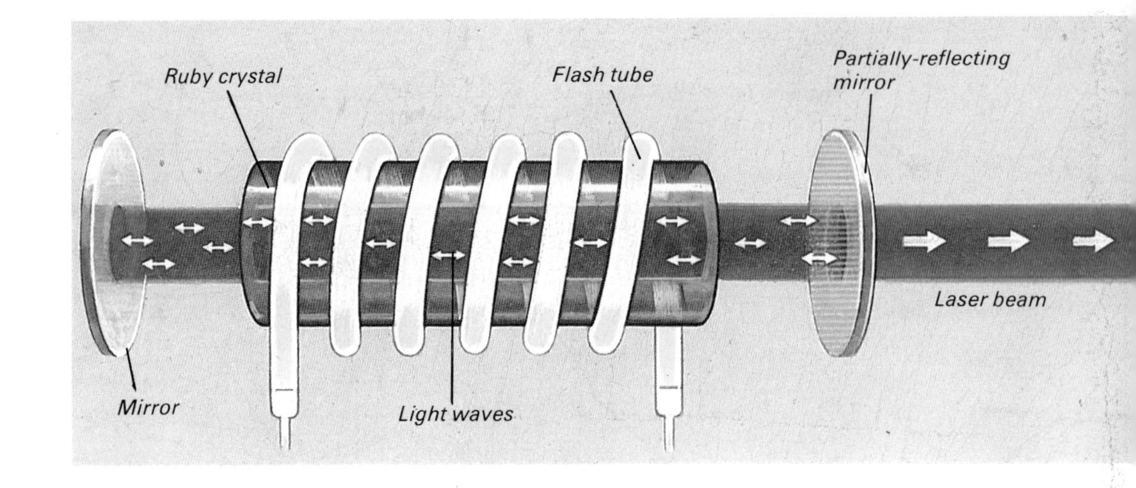

Above: A typical ruby laser consists of a flash tube around a ruby rod containing chromium atoms to produce laser light. Photons bouncing back and forth along the ruby rod stimulate more chromium atoms and a beam of coherent red light emerges via the partially reflecting mirror at one end of the ruby rod.

Dr. Theodore H. Maiman, who was largely responsible for the first laser at the Hughes Aircraft Company in California, is seen here with his ruby laser.

strip light and a laser. A strip light produces all its light through its sides; a laser produces its useful light in the form of a beam that emerges from one end of the laser material.

Inside the laser, photons are bouncing about in all directions. Some of these pass through the side walls of the lasing material and are lost. But in order to stimulate as many atoms as possible the laser keeps most of its light inside it. This is achieved by having two mirrors, one at each end of the material. One of these reflects all the light that hits it. The other is partially reflecting or has a transparent spot in the middle. Light travelling along the laser is bounced back and forth off these two mirrors. When the light passing up and down gains enough energy a small amount of laser light leaves the laser through the partially reflecting mirror.

It may seem surprising that, given all the talk of the power of lasers, only a small part of the light bouncing to and fro in the laser is let out. But this is essential to make efficient use of the excited atoms or molecules and so get the maximum output from the laser. Many lasers give out only a tiny part of the input energy as useful light. Even the most efficient lasers only return a quarter of the input energy as laser output. But the lost energy is a small price to pay for the unique properties of the laser beam.

Laser Beams are Special

When atoms or molecules are stimulated to produce photons by other photons, the light produced has a precisely defined wavelength and frequency. And, as all the waves are in step, the light is coherent.

The wavelength of laser light is also much more precisely defined because of the way in which the laser is constructed. Most of the light is reflected to and fro by the two mirrors. But this means that beams of light travelling in opposite directions meet in between the mirrors. If such beams are out of phase when they meet, they will interfere with each other. So, to ensure that the beams are in phase, the mirrors are placed an exact number of wavelengths apart.

This has another advantage. Most lasing materials in fact produce not one wavelength, but a tiny range of wavelengths. Just one of these wavelengths can be chosen with incredible precision by adjusting the distance between the mirrors. A coherent beam of light of precisely

known wavelength is a very useful tool, as it can be used to measure distances with unequalled accuracy.

A laser also produces unbelievably bright light. This is because the light is collected by the lasing material and amplified many times before it is allowed to leave. An ordinary light works at a few thousand degrees centigrade. To produce the brightness of a laser, an ordinary light would have to reach impossible temperatures—millions of degrees centigrade.

In producing such bright light the laser causes it to bounce to and fro along the laser material a large number of times. As a result, only light that is parallel to the axis of the material is involved in generating useful output. This means that the light comes out as an almost perfectly parallel beam, which can be projected over great distances with very little loss in intensity. Moreover, the quality of the beam makes it possible to focus the light down to a minute spot with far greater precision than light from an ordinary lamp can be focused.

A pencil-thin beam of light shoots from a gas laser (on right of picture).

Above left: Lasers do not have to be large. Here a tiny semiconductor laser (made from the same materials as a transistor) is sandwiched between two micromirrors near the tweezer points. When the laser is subjected to a high electric current, it produces pulses of light that last just one picosecond (one million millionth of a second).

Controlling the Output of Lasers

Many lasers give continuous, uninterrupted beams of light. Others can be made to give out light in short pulses.

Pulsed lasers are generally more powerful than continuous wave lasers. However, pulsed lasers can also be made from materials used for continuous wave lasing. In such cases the output can be controlled by ensuring that no light escapes until the lasing material is fully excited. Then the laser light is let out in a single, intense burst.

Light emitted from specially controlled pulsed lasers can be compressed into tiny fractions of a second (a few trillionths of a second). In this way they can be made to produce huge power levels. For an instant some lasers can produce a peak power greater than the output of all the electricity power stations in the world. At these high power levels the behaviour of materials is completely altered. Some lasers can make materials vaporize instantly and explode in a spectacular manner.

Less dramatic, but of equal importance, is the fact that at slightly lower power levels some materials can absorb laser light of one wavelength and then emit laser light of a lower wavelength. Converting laser light from one wavelength to another in this way is

An early continuous-wave argon laser being demonstrated in 1964. It had a power of eight watts and sent out a beam in the blue-green part of the spectrum.

Left: Some dyes can be made to absorb light of one wavelength and emit light of another wavelength. In a dye laser the dye is excited by an argon or krypton gas laser. A dye laser has the advantage that it can produce a number of different wavelengths within a fixed range. And it can be tuned to produce just one of these wavelengths at a time. The dye laser shown here produces a high power beam that can be finely tuned. And it is so stable that it can be made to produce the same frequency for hours at a time. It can be used for studying atoms and molecules and for holography.

known as *harmonic generation*. It produces wavelengths of precisely a half or a third of the original laser wavelength (they therefore have double or treble the frequency). This is useful in producing high power levels at high frequencies. Simple lasers cannot yet be made to do this efficiently.

The search for new lasers continues. The discovery of new materials that emit different wavelengths increases the number of ways of exploiting laser light. But the greatest effort is being directed into devising lasers that will produce high-frequency ultraviolet rays and X-rays. This is very difficult to achieve, but the potential rewards are great.

One form of control is of growing value. Scientists have discovered how to build tunable lasers. This can be done using two properties of certain materials. First, the energy levels in some atoms and molecules can be shifted in a magnetic field. So, if a laser material containing such atoms is put inside a magnet whose field strength can be adjusted, the wavelength of the laser can be altered over a wide range. But for any given magnetic field it still emits a precise wavelength. This is of great value in studying materials, as a tunable laser can be used to scan their properties with great accuracy.

Second, some liquids are capable of emitting light over a wide range of wavelengths. A tunable laser can be made using such a liquid in a laser cavity that can be adjusted—that is, the distance between the two mirrors can be altered. By adjusting the size of the laser cavity, a scientist can alter the wavelength of the light emitted by the laser as he requires.

Measurement by Laser

Laser light can be used to measure electric current. Laser light is polarized; that is it vibrates only in one plane. An electric current can change the angle of the plane of polarized light, the higher the current the greater the change in angle. Here, laser light is being used to measure the current flowing in a copper tube. The light is passed through an optical fibre (see page 50), which is wrapped around the tube. Instruments detect the plane of polarization of the laser light before and after it passes round the tube. The difference between the two readings gives a measure of the amount of current flowing through the copper tube.

The ability to make accurate measurements is essential to modern scientists and technologists. And the unique properties of laser light make it possible to measure things more accurately than ever before. Scientists have developed a variety of measurement techniques that exploit the different properties of laser light. Many of these techniques take advantage of the almost parallel beam, which makes laser light easy to direct with great accuracy at a selected target.

The fact that the wavelength and frequency of laser light can be defined with pinpoint precision is used in measuring distance and in several other measurement techniques. In fact, there are ways of locking a laser to such an exact frequency that, in future, the best way of defining all measurements of length may be in terms of laser light.

Because the light from a laser is coherent, considerable distances can be measured by counting the number of wavelengths from one point to another. This can be done to an accuracy of a tiny fraction of one wavelength. For measurement of even greater distances, the high power levels of lasers and their ability to form very short pulses provide ways of producing measurements of unequalled accuracy.

Tunable lasers, in which the wavelength of laser light can be adjusted, can be used to detect the atoms and molecules of pollutant chemicals. This provides many ways of measuring and studying pollution. At the same time, tunable lasers can be used to study the structure and other properties of molecules.

Modern physics has taken advantage of all these properties,

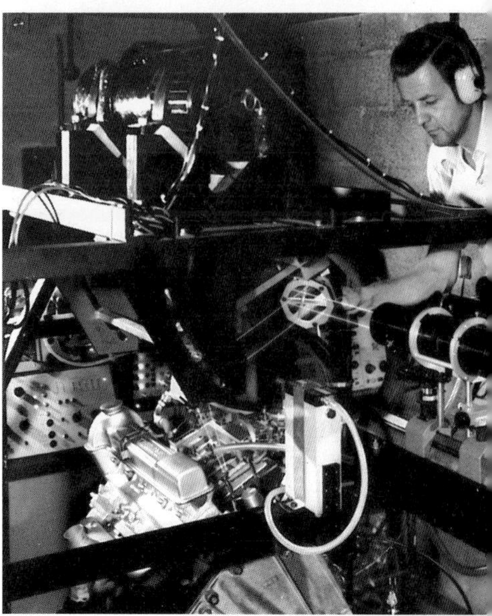

together with other discoveries, to make major advances. For example, lasers have been used to measure the speed of light with far greater accuracy than was previously believed possible.

Measuring Short Distances

The measurement of distance was one of the earliest and most obvious uses for lasers. The principle used in measuring short distances is known as *interference*. This is a property of light that occurs when light beams of identical wavelength meet. If the waves are in phase, they reinforce each other; the amplitude of the waves increases and therefore the brightness of the light increases. On the other hand, if the waves are totally out of phase (that is, when the peaks of one set coincide with the troughs of the other), they cancel each other out and no light can be seen. In between these situations, when the light waves are only partially out of phase, there is a range of brightness.

This property of light is used in a measuring device called an *interferometer*. This consists of a light source, a beam splitter and a fixed mirror. The light is split into two equal beams. One of these is directed along a short reference path to the fixed mirror. The other travels to a movable mirror. The two beams are reflected by the mirrors back to the beam splitter, where they recombine and pass to a detector.

To measure the distance between two points, the movable mirror is moved from one point to the other. As it moves, the brightness signal of the recombined beams varies as the two beams interfere with each other. The signal changes through one complete cycle for every half wavelength that the mirror travels.

Ordinary light can be used in an interferometer to measure short distances near the instrument. But beams of ordinary light are not coherent, even if they are produced from the same source, and the amount of interference between the beams is reduced to very low levels when the moving mirror is more than about one metre farther away from the beam splitter than the fixed mirror. Laser light, however, is fully coherent and produces very sharp signals. So an

Above left: Lasers are used to make measurements in a wide range of situations. Here, an instrument is measuring how fast the gases are moving just inside a small gas furnace. The instrument is known as a laser doppler anemometer. It operates by bouncing a beam of laser light off the moving gases and detecting the change in frequency of the returning beam.

Above: A laser doppler anemometer being used to measure gas movement in the manifold of a motor car engine.

interferometer using laser light can be used when the movable mirror is several kilometres away from the instrument.

The accuracy of a laser interferometer and the distances over which it can be used can produce very interesting results. It is possible to measure minute changes in distance between two points that are quite far apart. Using this ability, some exciting work is being done in measuring movements of the rocks in the Earth's crust. To do this an interferometer is mounted on a pillar which is sunk into bedrock. A mirror is then mounted on a second pillar anything from tens of metres to a few kilometres away. The instrument can measure any change in the distance between the two pillars to within less than a tenth of a micrometre (one ten millionth of a metre). However, to achieve this kind of accuracy, it is necessary to prevent the instrument being disturbed by air movements. So the whole apparatus is mounted in an evacuated container.

Lasers are now being used to study a wide range of rock movements. The Earth's crust is always being subjected to stresses and strains. These cause the crust to move continually, but most of the movements are too small for us to feel. One kind of movement is caused by the gravitational pull of the Sun and Moon. Minute tidal movements are produced in the crust in just the same way as the much greater tides that occur in the oceans. These tidal movements are being studied with previously unimagined precision.

At the same time scientists are investigating the intriguing possibility that the whole of the Earth's crust is being moved by gravity waves from outer space. These have not yet been detected for certain, but laser interferometers may well provide the best way of demonstrating their existence. If and when they are detected, scientists will have found the most elusive feature of Einstein's *General Theory of Relativity*—the explanation of gravity, which is the weakest and most mysterious of the forces of nature.

Other kinds of movements occur in the huge plates that make up the Earth's crust. These plates are floating on a semi-liquid layer

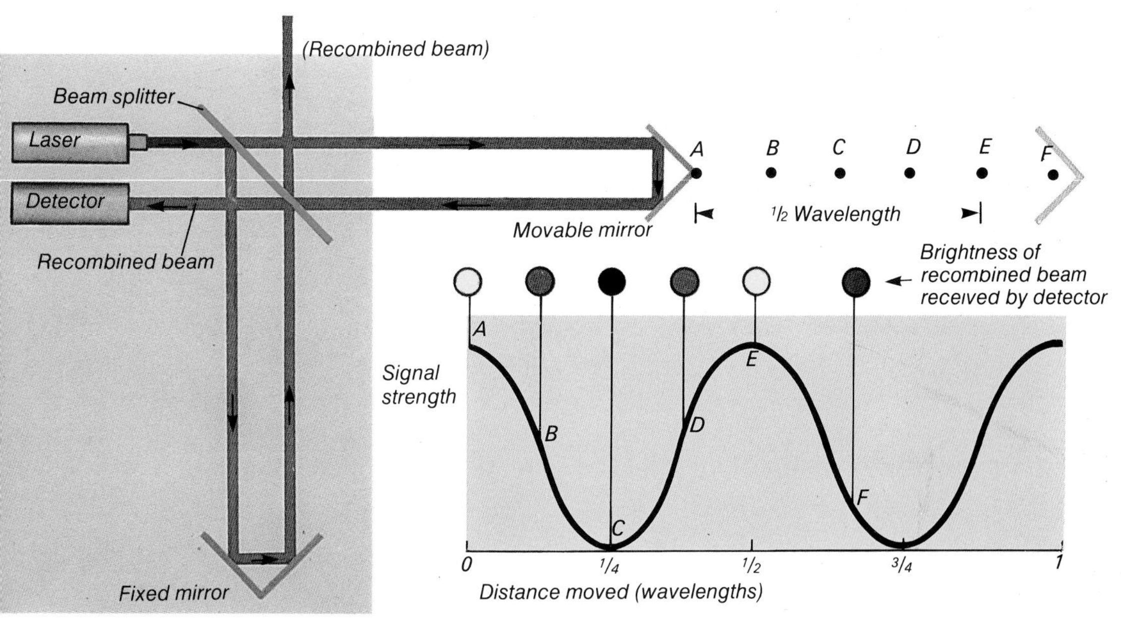

A laser interferometer is used to measure very small changes in distance. The beam from the laser is split into two identical beams by the beam splitter—a partially-reflecting mirror. One beam travels to a fixed mirror; the other travels to a movable mirror. These mirrors reflect the beams back to the beam splitter, where they combine and pass to the detector. As the movable mirror moves along the path of the beam the signal strength at the detector varies as shown on the graph—the light signal of the combined beam varies in brightness. The signal changes through one complete cycle as the mirror moves half a wavelength. In this diagram the movement of the mirror is greatly exaggerated for clarity— in fact it is very small indeed.

below. In some places these vast plates are moving apart, with new crustal material being added from beneath. In other places plates are moving together, with one plate being forced below the other. In yet other places plates are slowly sliding against each other. Such regions are known as *transform faults*—the San Andreas Fault in California is a typical example. The regions on each side of a fault are in continual motion, but the speed of this movement varies greatly. Sometimes the movement is enormous and an earthquake occurs. At other times there is no movement or only slight movements. Scientists believe that during these quiet periods enormous stresses build up and that large earthquakes occur when the stresses become too great and the pent up energy is released. A number of active fault regions are being studied using laser interferometers installed across them. It is hoped that such studies will enable scientists to predict future earthquakes.

Laser interferometers can also detect the tiny movements caused by earthquakes far away from the actual earthquake region. And, unlike other instruments, they can distinguish reliably between these movements and the vibrations caused by nuclear explosions thousands of kilometres away. In the future it should be possible to detect whether countries are breaking underground nuclear test ban treaties.

Above: Scientists using a laser interferometer to detect slight movements across the San Andreas fault in California. The aircraft is used to measure the temperature and humidity along the path of the laser beam. The movable mirror is several kilometres away, but the scientists are able to make distance measurements accurate to within a fraction of a millimetre.

Measuring Long Distances

The method of measuring long distances is less complicated. It simply involves observing the time taken for a pulse of light to make the round trip from a laser to an object and back to a receiver, which is usually beside the transmitting laser. Since the speed of light is known very accurately (see page 26), it is necessary only to observe precisely the time taken for a short pulse of light to make the journey.

This method also has the advantage that the distant object does not have to be a special reflector. In some cases a mirror is used, but it is often possible to rely on the natural reflectivity of the object itself. In this way lasers can be used to determine the position of distant, weakly reflecting targets, such as thin clouds or dust in the atmosphere.

Left: Another way of detecting movement along the San Andreas fault is to bounce laser beams off a satellite. The LAGEOS satellite is a geostationary satellite; that is it remains positioned over the fault. Laser beams from ground stations are reflected back and enable scientists to pinpoint the relative positions of the stations. Over a period of months or years any changes in these relative positions can be monitored and thus the north-south motion along the fault can be measured.

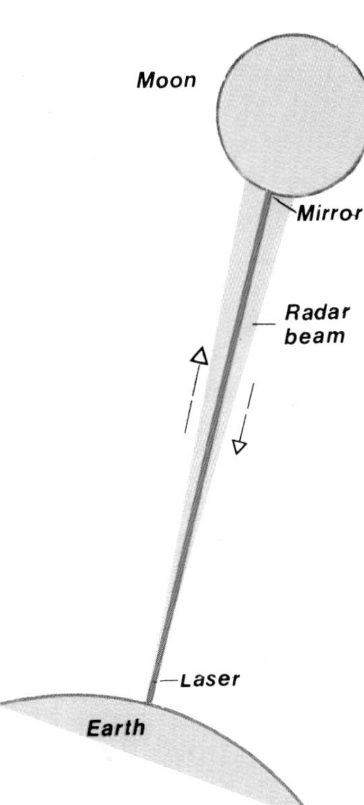

In July 1969 the Apollo 11 astronauts left a mirror on the Moon (above left). Since then this mirror has been used to make a number of measurements of the distance between the Earth and the Moon. Pulses of laser light are aimed at the mirror (above) and the distance is calculated from the time taken for the pulses to return to Earth. Radar pulses have also been bounced off the Moon, but, as the diagram shows, the radar beam spreads out over a vast area.

Laser range-finding systems can be used to measure very long distances indeed. For example, it has been possible to measure the distance from Earth to the Moon. When the Apollo 11 astronauts landed on the Moon they left a specially constructed mirror. Very short, intense pulses of light from a high power ruby laser are directed at the mirror through a large telescope system.

At this distance, of course, the fact that the laser beam is not quite parallel begins to have an effect. By the time the light pulse reaches the Moon it has spread over an area three kilometres wide. So only a tiny part of the light is reflected back to Earth by the mirror and an even more minute part (about one millionth of a millionth of the light that leaves the laser) is collected by the receiving system. Even so, this is enough to obtain precise measurements of the time taken for light to make the round trip of 800,000 kilometres to the Moon and back. After years of work, it is now possible to measure the distance of the Moon to an accuracy of five centimetres. This slight inaccuracy is not due to limitations in the performance of laser systems but to variations in the Earth's atmosphere, which affect the observations.

There are now five mirrors on the Moon, three left by the Americans and two put there by unmanned Russians space vehicles. These mirrors are being used to make a whole host of measurements. Scientists have studied the Moon's orbit, and the way in which both Earth and the Moon wobble as they move through space, with a precision that is totally impossible using other techniques. This has enabled them to make improved checks on Einstein's *General Theory of Relativity* and to show that, because of the energy dissipated in the tides, the Moon is actually moving away from the Earth at a rate of four centimetres a year.

Back on Earth the Moon ranging work is being used to measure changes in the distance between different parts of our planet. Using lasers, it is possible to detect precisely when, as the Moon orbits the Earth, it is closest to any given point on the Earth's surface. Any changes in such measurements indicate movements in Earth's crust. So far only a few measurements have been made; the distance between

Below: Laser binoculars enable a soldier to find the range of a target accurately and rapidly. All he has to do is aim the binoculars and press the trigger button. The range then automatically appears in the left hand eyepiece.

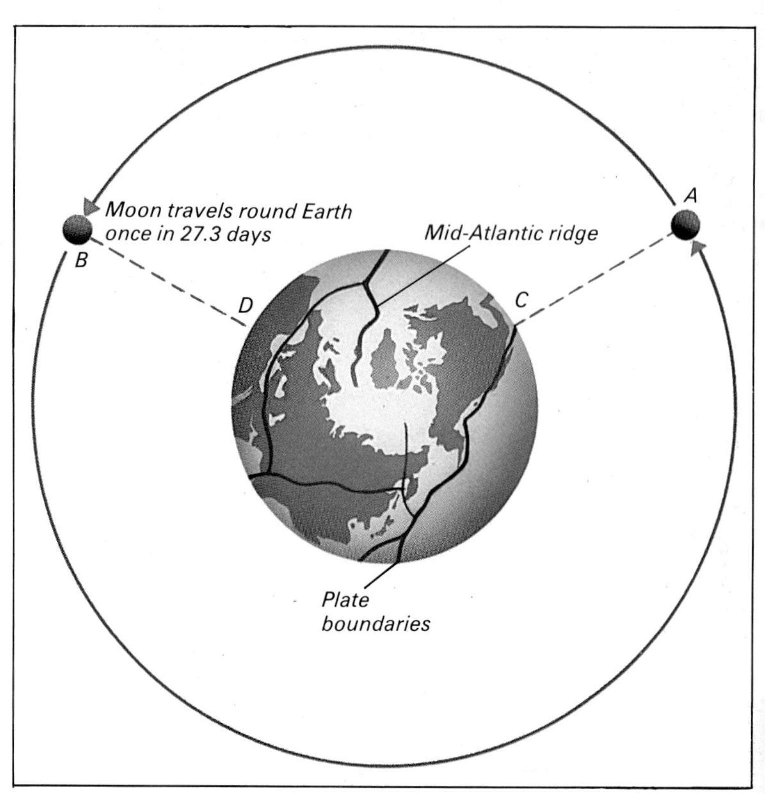

Below: Laser binoculars. When the trigger button is pressed, a pulse of high intensity laser light is directed at the target. A small part of the light pulse is fed to the receiver, which starts the electronic timer. The light returning from the target stops the timer and the range is computed from the time recorded.

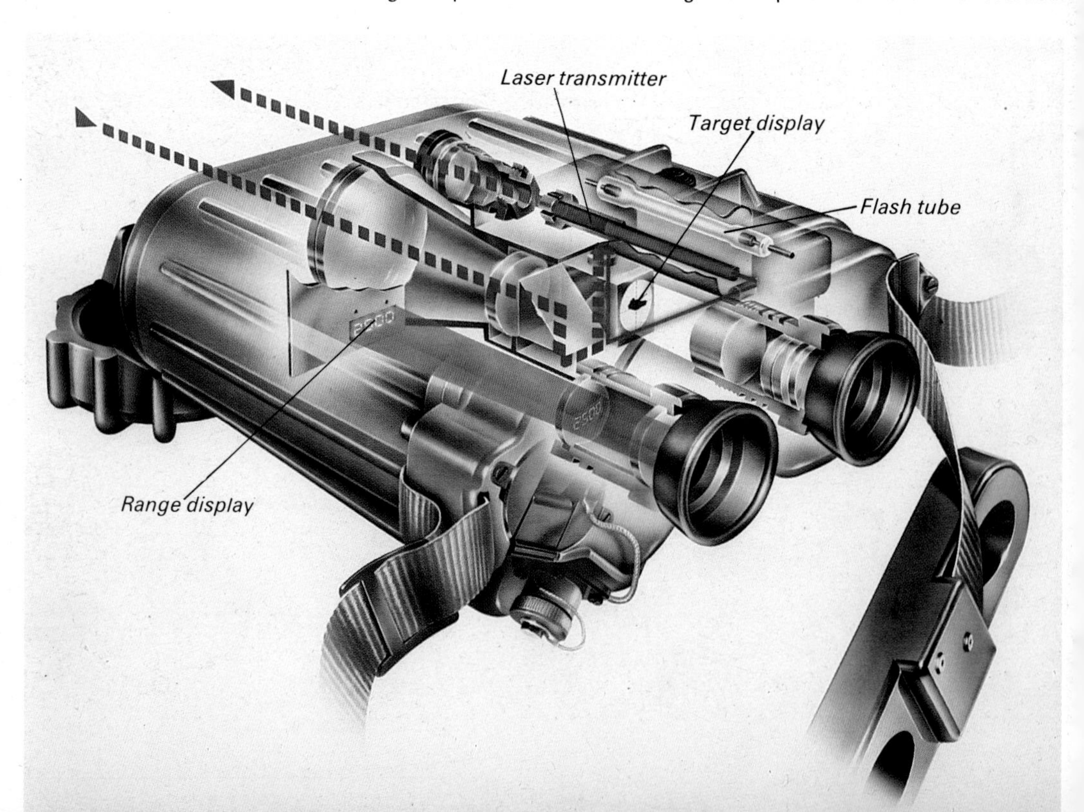

points in Texas and The Crimea is now known to an accuracy of one metre. In future scientists may be able to observe the movements of the plates that make up the Earth's crust. The continents, which rest on some of these plates, move with them. However, even a small movement takes a long time (the Atlantic Ocean is widening at only about two centimetres a year). Using lasers it will be possible to detect the movements of continents (known as *continental drift*) with an accuracy that is not possible by other methods.

Left: Moon ranging by laser can be used to help measure the rate of continental drift on Earth. For example, it should be possible to measure the rate at which the mid-Atlantic ridge is widening the gap between America and Europe. From the time taken for the Moon to travel from A to B and the laser-measured distances CA and DB it will be possible to calculate the distance CD on Earth very accurately. As the Atlantic Ocean widens, the time taken for the Moon to travel from A to B will increase and so the increased distance between C and D can be calculated.

The principle of finding the distance of remote objects by bouncing pulses of laser light off them has also been used in much smaller systems. Hand-held binoculars containing a small battery-powered crystal laser have been developed for military range-finding. Such binoculars can measure the distance of a target—a tank, for example—with an accuracy of five metres at a range of 10 kilometres. Similar but more complicated systems can be used as rapid-response altimeters in aircraft that have to fly low to avoid detection by enemy radar.

Alignment by Lasers

The use of lasers to line things up is just about the simplest way of using their properties. Since light travels in straight lines and lasers can produce such narrow beams, they provide an almost perfect way

Laser beams are widely used in the construction industry. Here, a laser beam is being used to check the alignment of a tunnel.

of aligning objects. Laser beams need no support and no additional reference points. So to build a bridge, road or any large construction all that needs to be done is to fix a laser at one point and shine a visible beam at another point. Then, providing nothing blocks the beam, it can be used to help the construction workers decide precisely where to place the various parts of the construction.

This is why one of the earliest and largest markets for simple lasers developed in the construction industry. Today, lasers are used to help speed up work in all kinds of major constructions, from roads and tunnels to bridges and high-rise buildings.

The Speed of Light

The laser has led to a huge improvement in the measurement of the speed of light. As mentioned previously, this speed (c) is constant and is defined in terms of wavelength (λ) and frequency (f) by the simple formula $c = \lambda \times f$. This means that to arrive at a figure for the speed of light all that is needed is to measure the wavelength and frequency of a very stable laser.

However, this process is not as simple as it sounds. This is because the frequency and wavelength of the laser have to be compared with established standards of frequency and length. These standards are in very different parts of the electromagnetic spectrum and bridging this gap and making full use of the advantages of stable lasers is no easy task.

The standard of frequency (and incidentally of time) is defined by a

Below: The caesium atomic clock is regulated by the vibration of atoms. It operates at a frequency of about 9000 million cycles per second and is accurate to 10 million millionth of a second.

The forerunner of the laser gyroscope. This ruby laser is under test and the beam is being reflected by mirrors into a closed loop. Tests like this led to the development of the first laser gyroscope in 1962.

device known as a caesium atomic clock. This has a frequency of about 9,000 million cycles per second (9,000 megahertz), which corresponds to light of a wavelength of about 33 millimetres. The frequency of this clock is the most accurate of all physical standards and can be used with an accuracy of a few parts in 10 million million.

Before the advent of lasers the best way to set the standard for length was in terms of the wavelength of visible light (wavelength 0.61 micrometres) produced by a special lamp filled with a gas called krypton. But because of the limitations of ordinary light (see page 14) this wavelength can be measured only with an accuracy of a few parts in a billion. The most accurate measurement of the speed of light was achieved by measuring the wavelength and frequency of a very stable microwave source with a wavelength of about 4 millimetres. However, the accuracy with which the wavelength of the microwave source could be measured was limited, because of the great difference between the wavelength of microwaves and the visible light produced by the krypton lamp. Hence, the speed of light could only measured to an accuracy of a few parts in 10 million.

Laser light has two great advantages in making such measurements. First, the wavelength of a stable laser can be controlled with greater precision than the krypton standard. Second, the higher power levels make it possible to generate harmonics of, or multiply up, (see page 17), low frequency lasers and so make exact measurements of higher frequency lasers. This means it has been possible to establish a standard for length using a stable laser which has a wavelength close to the krypton standard and also to use a string of lasers to bridge the gap between the frequency and wavelength standards. At the National Physical Laboratory at Teddington in London four lasers have been used by scientists to make a series of "stepping stone" measurements.

First, a caesium atomic clock frequency was multiplied up to measure the frequency of a stable hydrogen cyanide laser—a frequency of 891,000 megahertz (wavelength 337 micrometres). Then the signal from this laser was multiplied up to measure the frequency

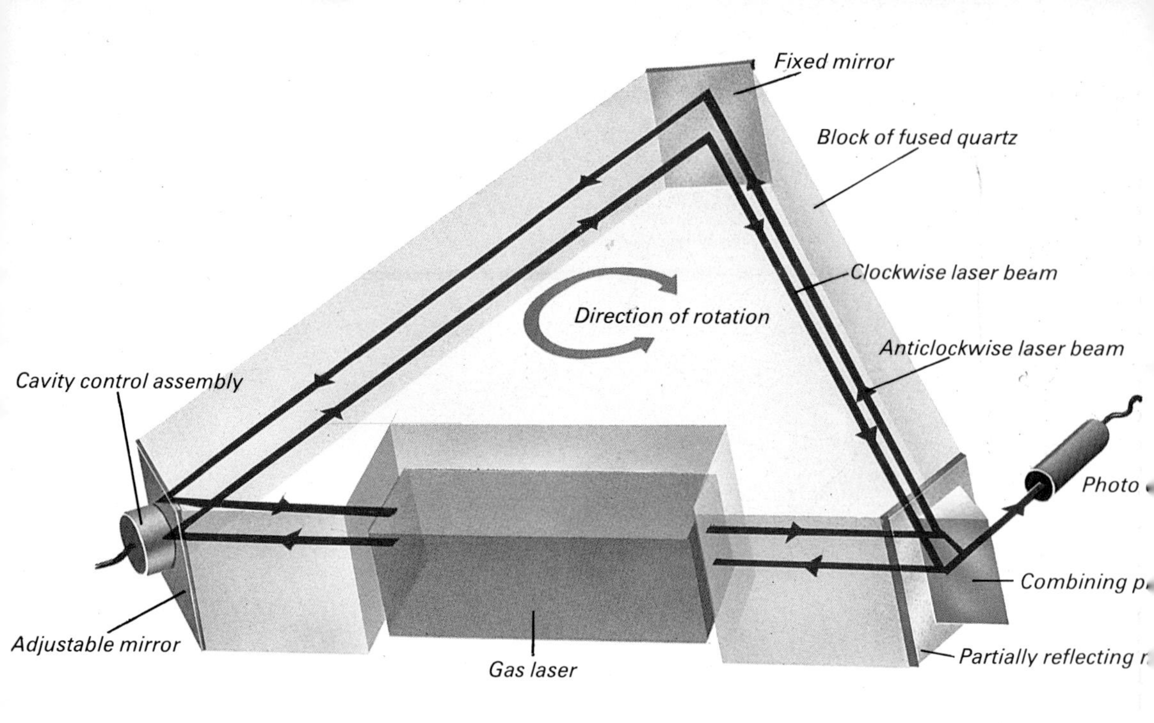

Above: A single axis laser gyroscope. Laser light is produced from both ends of the gas discharge tube. The two beams travel round the gyroscope in opposite directions. When the gyroscope rotates on its axis, the frequency of one beam increases, while the frequency of the other beam decreases. The beams are combined by the prism and the difference between their frequencies is detected by the photo diode.

of a stable water vapour laser—10.7 million megahertz (wavelength 28 micrometres). This laser was, in turn, used in the same process to measure the frequency of a stable carbon dioxide laser — just over 32 million megahertz.

At the same time a krypton standard light was used to measure the wavelength of the red light produced by a helium-neon laser—a wavelength of 0.63 micrometres. This laser, in its turn, was used to

Right: A three-axis laser gyroscope for use in missiles and torpedoes.

Far right: A model of the light pathways in a three-axis laser gyroscope.

measure the wavelength of the carbon dioxide laser—9.3 micrometres.

So, at last, the scientists had reached a point where they had accurate measurements of both the frequency and the wavelength of one laser—the carbon dioxide laser. Using these measurements they calculated the speed of light as being 299,792,459 metres per second. This figure agrees with the value produced in the United States using a different set of lasers. It is a hundred times more accurate than the figures that were previously available.

But why is it necessary to measure the speed of light to nine or more figures? This is far more accurate than is required for most ordinary calculations. The point is that, in modern physics, scientists depend on being able to make very precise measurements and calculations. Only by using such exact figures is it possible to expose the flaws or limitations of current theories. Once these have been shown up, it is then possible to produce new or improved theories, which can lead to new discoveries.

Laser Gyroscopes

Possibly the most extraordinary demonstrations of how stable the frequency of a laser can be is to use one as a gyroscope. This is done by building a closed laser in which the laser light goes round and round in a loop, like the flywheel of a mechanical gyroscope. Both triangular and rectangular versions have been built, with mirrors at each of the corners. To prevent changes in temperature from affecting the system, the loop is machined out of a single block of fused quartz. The light travelling round the loop locks onto a single frequency with amazing precision. And this fact can be used to detect movements of the laser. If the loop moves in any direction along its own plane, the acceleration causes the light waves to bunch up or stretch and so the wavelength and frequency change very slightly. Such tiny changes in frequency can be detected and used to calculate the acceleration of the laser.

By using three such laser gyroscopes to measure the acceleration along three perpendicular axes it is possible to build an inertial guidance system. Such systems are more accurate than mechanical gyroscope inertial guidance systems and they will be cheaper to build. They are therefore likely to replace conventional equipment in aircraft and ships of the future.

Measuring the Environment

The simplest forms of environmental studies use lasers as range finders. The study of smoke plumes, clouds or even volcanic dust high in the atmosphere can easily be carried out using laser systems. Pulses of light from a laser are reflected off the target and collected by a receiver and detector. Using sophisticated electronics it is possible to build up a complete picture of complicated smoke plumes from tall chimney stacks. After the eruption of the Mount St Helens volcano in Washington in 1980 the movement of the great dust cloud was tracked and studied using laser systems.

An even more detailed picture of clouds of pollution can be built up using two lasers. If they have slightly different frequencies a great deal more information can be obtained from the reflected signals. It is possible to measure not only the concentration of pollution or dust in a cloud but also the speed at which it travels and its internal movements.

Below: The absorption spectrum of sodium vapour. When the spectrum is recorded photographically, two dark lines appear in the orange band (top). Sodium vapour absorbs light energy at these two wavelengths, which correspond to the energy levels of its atoms. When the spectrum is recorded as a graph these lines show up as troughs (bottom). The separation of the lines has been exaggerated for the sake of clarity.

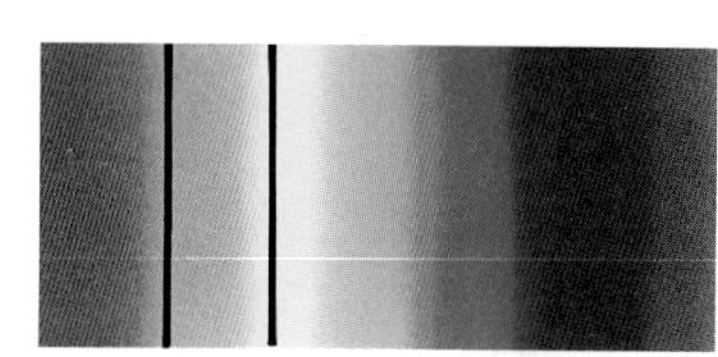

Long Wavelength *Short wavelength*

Amount of light energy absorbed

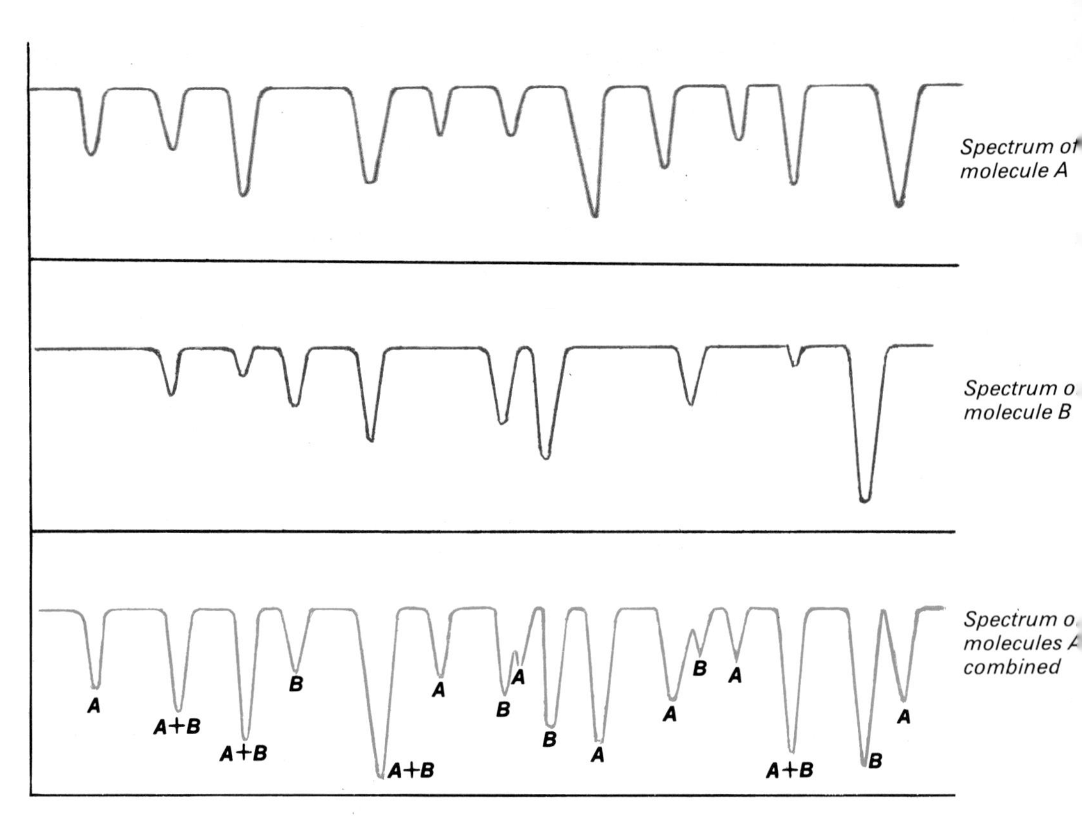

When the absorption spectra of individual substances are known, then it is possible to pick them out in the absorption spectrum of a mixture of the substances. For example, shown here are the absorption spectra of two substances (molecule A and molecule B). In the spectrum of the mixture the features of both substances can be seen. In some parts of the spectrum they remain the same; in other parts features combine to give deeper troughs.

This is of value in both pollution studies and meteorology.

The use of lasers in measuring pollution is not limited to range-finding work. Lasers can be tuned into the exact features of particular atoms and molecules. This means that they are well suited for making measurements of small quantities of one substance, even when many others are present. They can detect tiny traces of one material while completely ignoring everything else. This makes them valuable tools for detecting pollution of air and water.

Lasers can detect different substances by using the fact that each type of or molecule absorbs light in a particular way. In fact every substance has its own "light fingerprint" by which it can be immediately recognized. Such a "fingerprint" is known as an *absorption spectrum.*

An ordinary spectrum (like a rainbow) is produced when white light is split up into its various wavelengths—each wavelength is seen as a different colour. But when light passes through a substance, some wavelengths excite the atoms or molecules in the substance to higher energy levels. Some of the light energy at these wavelengths is therefore "used up" or absorbed by the material. If the remaining light that comes out of the material is then split up into its spectrum, certain wavelengths will be considerably less bright. This, then, is the "fingerprint" or absorption spectrum of the material.

An absorption spectrum can be studied photographically, in which case the absorbed wavelengths show up as dark lines among the colours. On the other hand, the spectrum can be measured by a detector which records the amount of light at each wavelength. The output of the detector can be recorded in the form of a graph, in which case the absorbed wavelengths appear as troughs.

If the absorption spectra of substances are known, it is then possible to detect them in a mixture of substances by shining light at the mixture and analysing the absorption spectrum obtained. Ordinary light can be used to do this, but with the low powers available it is not possible to distinguish all the features of the absorption spectrum. This is a major drawback, because it is then difficult to detect and recognize the different absorption spectra of all the chemicals present in the mixture. But with the high power and precise frequencies of laser light all the features of the absorption spectrum stand out clearly.

A tunable laser is used to scan a mixture of unknown substances with all its frequencies in turn. A detector records the amount and wavelength of any light that is absorbed. Scientists have now developed a large number of tunable lasers. Together they cover a wide range of useful wavelengths—0.4 to 40 micrometres. This wide range has an additional advantage in that it includes the wavelengths absorbed by the vibrational energy levels of molecules (see page 13). So it is possible to choose a laser to study any one of a wide variety of atmospheric pollutants and to concentrate on the most individual aspects of each pollutant.

Lasers are therefore particularly useful when studying a very complex pollution problem. A good example of this is the formation of photochemical smog, which occurs when, in the presence of strong sunlight, the chemicals in the exhaust from petrol engines react with each other and the oxygen in the air and then become attached to particles of dust or droplets of water. Photochemical smog is particularly troublesome in the Los Angeles basin, where the surrounding hills help to trap the smog.

One way of studying such pollution is to draw a sample of the polluted air into a special measuring cell and then scan it with a tunable laser. From the absorption features that are recorded it is possible to determine which pollutants are present and in what quantities. By taking samples at different times and in different places scientists can build up a picture of the way in which smog develops and changes.

Another way of looking at atmospheric pollution is to measure the absorption that occurs when a laser beam is directed over a long path through the atmosphere. The beam is aimed at a mirror some distance away and is reflected back to a receiver. It is then possible to measure the total absorption at each wavelength over this path. Using this method scientists have less control over the sample, but they do get a more direct measure of what is going on in the atmosphere as it actually happens.

A tunable laser has also been used to study the same sort of chemical changes that occur high in the atmosphere. A ballon-borne laser system and sampling cell was flown to an altitude of 30 kilometres. Measurements were then made during both day and night of the changes in concentration of certain gases that exist in tiny amounts in the upper atmosphere. These measurements were much more accurate than those obtained by other methods.

This may not seem a very important experiment, but these gases are actually vital to life of Earth. In particular, the gas known as ozone plays the vital role of screening out certain ultra-violet radiation that would be lethal to many forms of life. There have been widespread fears that pollution from high-flying aircraft or even gases from aerosol cans might destroy this ozone layer. Laser studies of the trace gases involved in the chemical creation and destruction of the ozone layer may help us to understand this important issue.

Although absorption measurements are the most obvious way of studying atmospheric pollutants, other techniques can also be used. One of these uses the fact that when molecules are illuminated by a strong beam of light, they re-emit a tiny amount of light, at a different wavelength which can be detected by a sensitive instrument. This technique does not give the same accuracy as absorption methods, but it can be used, together with range-finding apparatus, to study the dispersal of various gaseous pollutants in smoke plumes. It can also be used to measure combustion processes in inaccessible places, such as in furnaces and gas turbines.

Above: A laser beam being used to measure cloud height at an airport.

There is still another way of using lasers to detect distant molecules. This takes advantage of the fact that in the mid- and far-infra-red every molecule naturally emits its own frequencies. These correspond to low energy rotational and vibrational motions which are excited at normal temperatures. Such radiation is normally very hard to detect, but it can be greatly enhanced by mixing it with the same frequency from a laser. This is known as *heterodyne detection* and is so sensitive that equipment on Earth has been used to measure the amounts of carbon dioxide on Mars and Venus and of ammonia on Jupiter. One extraordinary result of this work has been to show that high in the atmosphere of Mars carbon dioxide is so excited that it is capable of laser action. In effect it is a natural laser. It is also being developed for remote sensing of trace gases in the Earth's atmosphere and may be used in balloons or satellites.

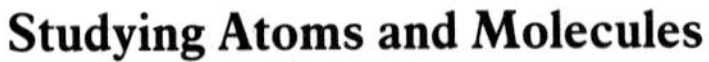

Studying Atoms and Molecules

Using tunable lasers to examine the unique absorption spectra of materials offers exciting possibilities in studying the structure of atoms and molecules. When the absorption lines (dark lines) of an absorption spectrum are recorded as troughs on a graph, the depth of each trough shows how much light energy is used up at a particular wavelength as the atoms or molecules are excited from one energy level up to the next. Therefore, the sizes and shapes of the troughs and the spacing between them tell scientists many things about the energy levels and hence the structure of the atoms or molecules. In addition, the absorption spectrum may change when the substance is subjected to different temperatures and pressures. Electrical or magnetic fields may also cause changes in the spectrum. Such changes tell scientists much about the physical and electrical properties of the atoms and molecules and the way in which they interact with one another.

Absorption spectra can be produced and studied using ordinary light and an instrument called a spectrometer. Much work has been done on the absorption spectra of many materials in this way. But when ordinary light is split into its rainbow spectrum the amount of light that occurs in any one part of the spectrum is very small—too small to allow scientists to detect very narrow absorption lines.

A laser however, produces much more light over a much narrower range of wavelengths. And this range is so small that it is possible to detect the narrowest absorption line. Using a tunable laser, therefore, scientists can tune into and scan all the features of an absorption spectrum with extraordinary precision. As more and more substances are studied in this way, scientists will gain a completely new insight into the structure and behaviour of atoms and molecules.

A New Chemistry

Powerful lasers offer the possibility of a completely new kind of chemistry. For the first time it may be possible to feed controlled amounts of energy into a chemical reaction and control the way in which individual molecules behave.

To some extent chemists can already control chemical reactions. Many reactions can be speeded up by raising the temperature or pressure. Other reactions can be helped by the presence of a *catalyst*—an ingredient that helps cause the reaction but is not itself used up in the reaction. But all the controls so far available have one major limitation—in any reaction all the molecules present are acted upon at the same time.

Lasers, however, offer new prospects. Because the light from a laser has a precise wavelength and frequency, it can be used to pump energy into particular molecules and in a very specific way. This has opened up exciting new possibilities for the chemistry of the future.

The three isotopes of hydrogen.

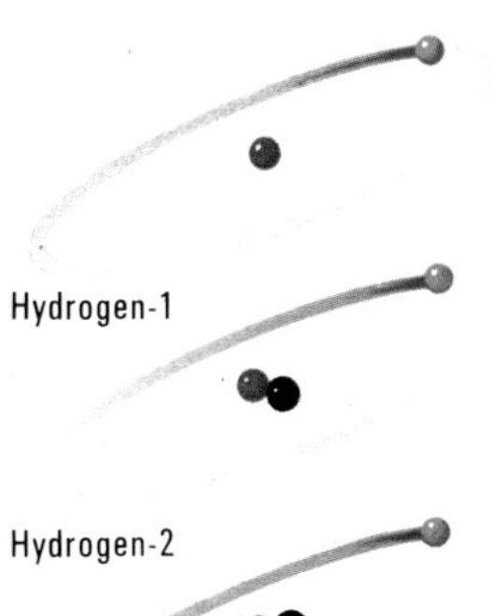

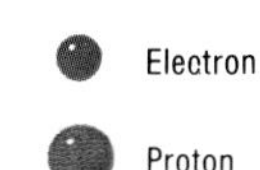

Separating Isotopes

All elements are made up of atoms. At the centre of an atom there is a nucleus. This contains positively charged particles called *protons* and uncharged particles of almost exactly the same mass called *neutrons*. Orbiting around the nucleus are negatively charged, much lighter particles called *electrons*. The mass of an atom is made up almost entirely of the protons and neutrons it contains.

In any atom the number of protons and electrons are the same and their charges balance each other out. There are over 100 different known elements and each one has a specific number of protons in the nuclei of its atoms. For example, a hydrogen atom always has one proton, a carbon atom has six and a chlorine atom has 17. But the atoms of an element do not necessarily all have the same number of neutrons. For example, chlorine atoms can have either 18 or 20 neutrons. These two forms of chlorine are known as chlorine-35 (17 protons and 18 neutrons) and chlorine-37 respectively—the numbers are known as the *mass numbers* and are arrived at by adding up the numbers of protons and neutrons in the nucleus. They have identical chemical properties, as the neutrons have no electrical charge and there are still 17 protons and 17 electrons in the atoms. But chlorine-37 has a slightly greater mass; that is, it is slightly heavier than chlorine-35.

Most elements exist in two or more forms and different forms of the same element are known as *isotopes*. Because they have the same chemical properties, isotopes cannot be separated by chemical means. Instead they have to be separated by lengthy and expensive techniques that depend on their different masses. One way is to allow a mixture of isotopes, in the form of a hot gas, to diffuse up and down a huge system of porous columns. The different isotopes diffuse through the porous material at different rates. More recently, centrifuges have been used to spin substances at very high speeds and so separate the isotopes.

Usually, the fact that an element is a mixture of isotopes is

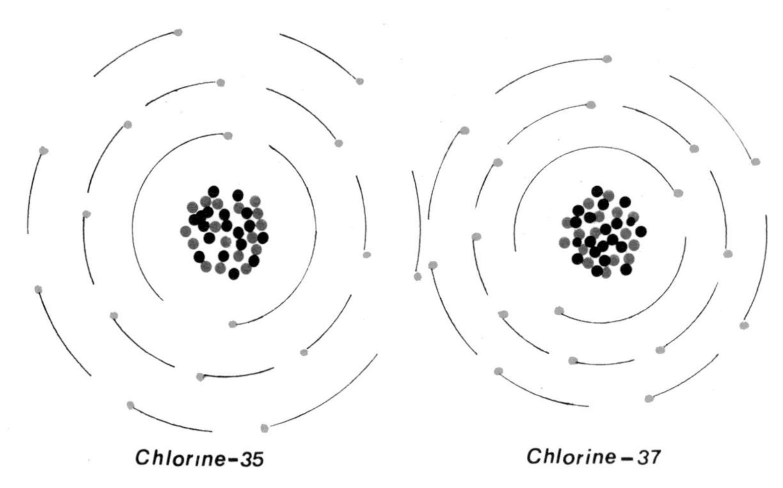

The two isotopes of chlorine.

unimportant. For example, although iron is a mixture of four main isotopes, it is not necessary to separate them. But in some cases one isotope is considerably more important than the others. This is particularly true of radioactive isotopes, or *radioisotopes*.

Probably the best known radioisotope is uranium-235, which can be used to generate nuclear power and to make atomic bombs. But nearly all uranium is uranium-238, which does not undergo nuclear fission; only 0.7 per cent of a piece of uranium is uranium-235 Nuclear fuel does not have to consist entirely of pure uranium-235, but it usually has to be *enriched* by separating out a large proportion of the uranium-238. Until now this separation process has required huge plants that use large amounts of energy. But lasers offer the prospect of a cheaper, more efficient way of separating isotopes.

Because different isotopes have different weights, they have different spacings between the energy levels of their atoms. The energy levels correspond to the orbits of the electrons, and heavier nuclei tend to hold their orbiting electrons closer to them. When isotopes are combined with another element to form a chemical compound, the molecules (groups of atoms) of the compound will also have different energy levels, according to which isotope is present in each molecule. Therefore it is possible to choose laser light of a wavelength that corresponds precisely with the energy level separation of the molecules containing one particular isotope. This laser light will excite only these molecules, which then heat up. All the remaining molecules, which contain other isotopes, remain unexcited and unheated.

The next problem is to single out the heated molecules. One way of doing this is to make them react with the molecules of a different chemical compound. By selecting the right compound it is possible to ensure that only the heated molecules react, producing a new chemical that contains the required isotope. This isotope is then in a different chemical from all the other isotopes and can therefore be separated chemically.

This method has been used to separate chlorine-35 and chlorine-37. When a compound called iodine monochloride is subjected to a laser beam of a particular wavelength, only the molecules containing chlorine-37 are excited. If this is done in the presence of a chemical called bromobenzene, only the excited molecules react with the

Below: Isotopes have different masses and hence the separation of their energy levels is different. Each energy level corresponds to only one particular wavelength. This means that in a mixture of isotopes it is possible to excite just one isotope by bombarding the mixture with laser light of the right wavelength. In the diagram below, the red lines show the energy levels of isotope A, the black lines the energy levels of isotope B. Only A isotope will absorb laser light of the left-hand frequency.

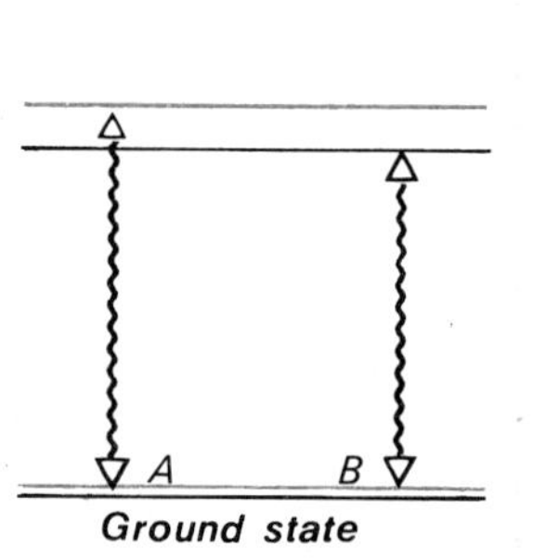

KEY TO ATOMS
(right)

- Carbon
- 35 Chlorine-35
- 37 Chlorine-37
- Bromine
- Iodine

bromobenzene to form chlorobenzene. This compound, which contains the atoms of chlorine-37, can then be separated out by chemical means.

There is another, simpler way of exploiting the fact that molecules can be selectively excited by laser light. When an extremely intense laser beam is used, those molecules that can absorb the light soak up so much energy that they simply fall apart. This effect was first discovered in a chemical compound called sulphur hexafluoride (SF_6). When a molecule of this compound is subjected to intense light from a carbon dioxide laser, it loses a fluorine atom and is converted to sulphur pentafluoride (SF_5), which is unstable and breaks down. By using different wavelengths of laser light it is possible to separate the sulphur isotopes sulphur-32 and sulphur-34. When light of wavelength of 10.6 micrometres is used, molecules containing sulphur-32 are converted (see diagram on page 36). With a wavelength of 10.82 micrometres molecules of sulphur-34 are converted.

A similar process can be used to separate uranium isotopes. In this case very high frequency lasers are used to excite uranium atoms and knock an electron out of them. This gives the atoms an electric charge and they can then be collected by means of an electric field. One laser by itself does not have sufficient power to do this. But by combining

Separating chlorine isotopes. **1.** The two isotopes, in the form of iodine monochloride, are mixed with bromobenzene. **2.** The mixture is exposed to laser light that excites only the chlorine-37 atoms. **3.** The iodine monochloride molecules that contain chlorine-37 atoms break down and iodine is formed. The excited chlorine-37 atoms then react with the bromobenzene to form chlorobenzene and bromine. **4.** The chlorobenzene (which contains only chlorine-37 atoms) is separated chemically. The chlorine-35 atoms remain behind, still contained in iodine monochloride molecules.

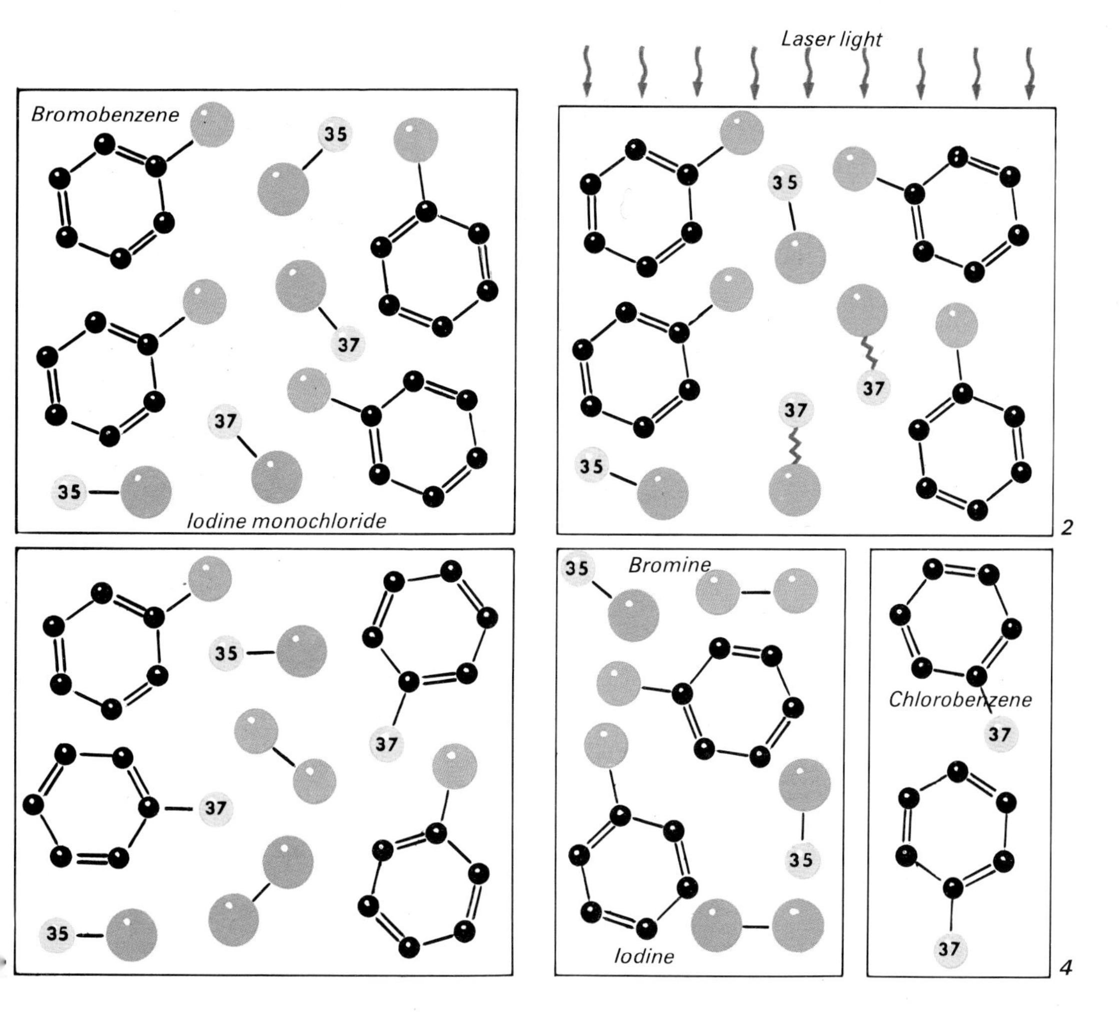

the powers of two ultra-violet lasers it is possible to knock electrons from the uranium-235 atoms in a beam of atoms that contains both isotopes. The uranium-235 atoms are then separated from the more common uranium-238 atoms by collecting them on an electrically charged plate.

An even simpler way of selecting certain isotopes uses the fact that when an atom or molecule absorbs a photon (wave packet of light energy) it recoils slightly. So if a beam of atoms or molecules is bombarded by laser light which is absorbed by the atoms of only one kind of isotope, then those atoms (or the molecules that contain them) can be guided away from the others and collected separately.

Almost all the work that has been done so far in laser chemistry has been with gases. This is because gases have the most clearly defined energy levels. So it is possible to choose a laser wavelength that will match exactly the energy level separations of only one type of gas molecule, leaving other molecules unaffected. Solids, on the other hand, have much broader energy levels and in liquids they are blurred out into great wide bands. This means that lasers cannot generally be used for sorting out the molecules of solids and liquids.

There are, however, exceptions. In particular, some crystals have very sharply defined energy levels, especially at very low temperatures. With crystals containing different isotopes of more than one element, it has been possible to excite one group of isotopes and break up the molecules in which they are contained. This has been done with a crystalline compound called tetrazine, which contains carbon and nitrogen. Tetrazine molecules containing the common isotopes carbon-12 or nitrogen-14 have energy levels that are different from those of molecules containing the much rarer carbon-13 and nitrogen-15 isotopes. By subjecting tetrazine to laser light for a few minutes it has been possible to break up the molecules containing carbon-12 or nitrogen-14. The resulting crystals of tetrazine are enriched with over 1,000 times the amount of carbon-13 and nitrogen-15.

Laser 'Catalysts'

Another possibility of great interest to chemists is the use of lasers in place of catalysts. It has been found that lasers can set off certain chemical reactions very efficiently. In doing so they act like catalysts, steering the reaction in the desired direction and triggering a chain reaction, using very little energy in the process. This can occur if a laser excites a vibrational energy level in a molecule, which then makes it much more likely to react with a nearby different molecule. The energy released in this reaction can then trigger further reactions.

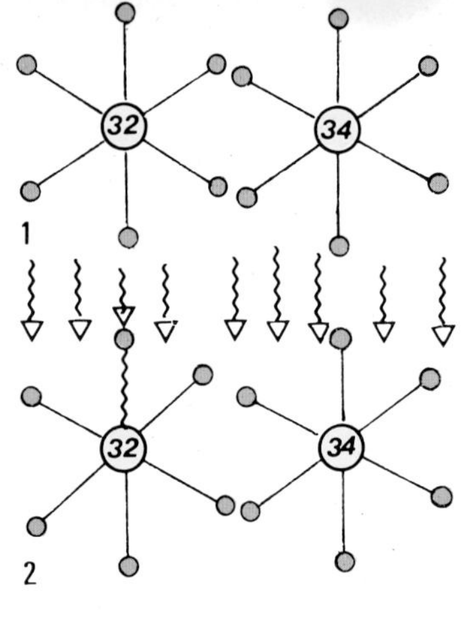

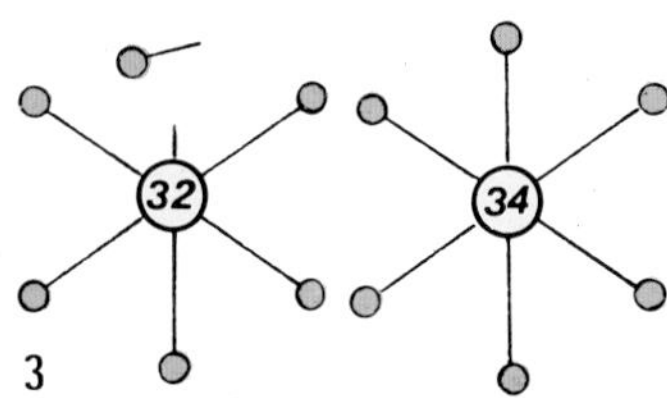

Above: Separating sulphur isotopes. **1.** Both isotopes are present in molecules of sulphur hexafluoride. **2.** When the sulphur hexafluoride is exposed to laser light of wavelength 10.6 micrometres, only sulphur-32 atoms are excited. **3.** The molecules containing sulphur-32 atoms lose a fluorine atom. The sulphur-32 atoms are now contained in molecules of sulphur pentafluoride, which breaks down and the resulting chemical can be separated chemically.
Below: Separating uranium isotopes. The beam of uranium atoms is subjected to ultra-violet light from two lasers. The uranium-235 atoms each lose an electron and become positively charged. Thus they are attracted to the negatively charged plate, whereas the uranium-238 atoms are not affected.

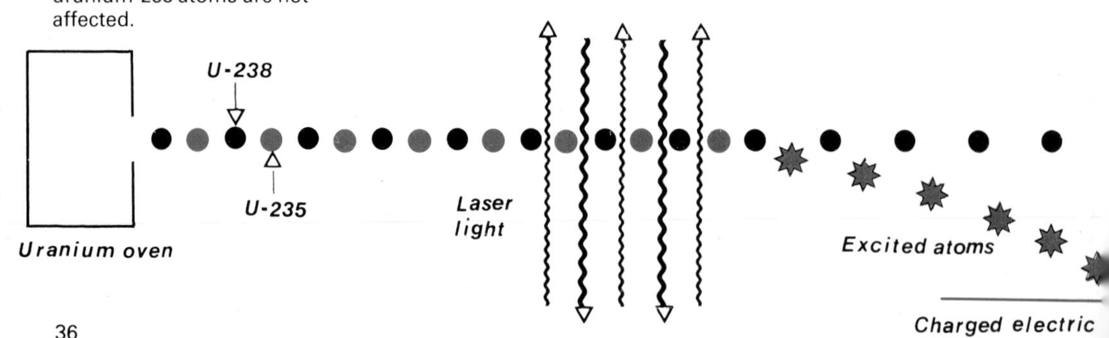

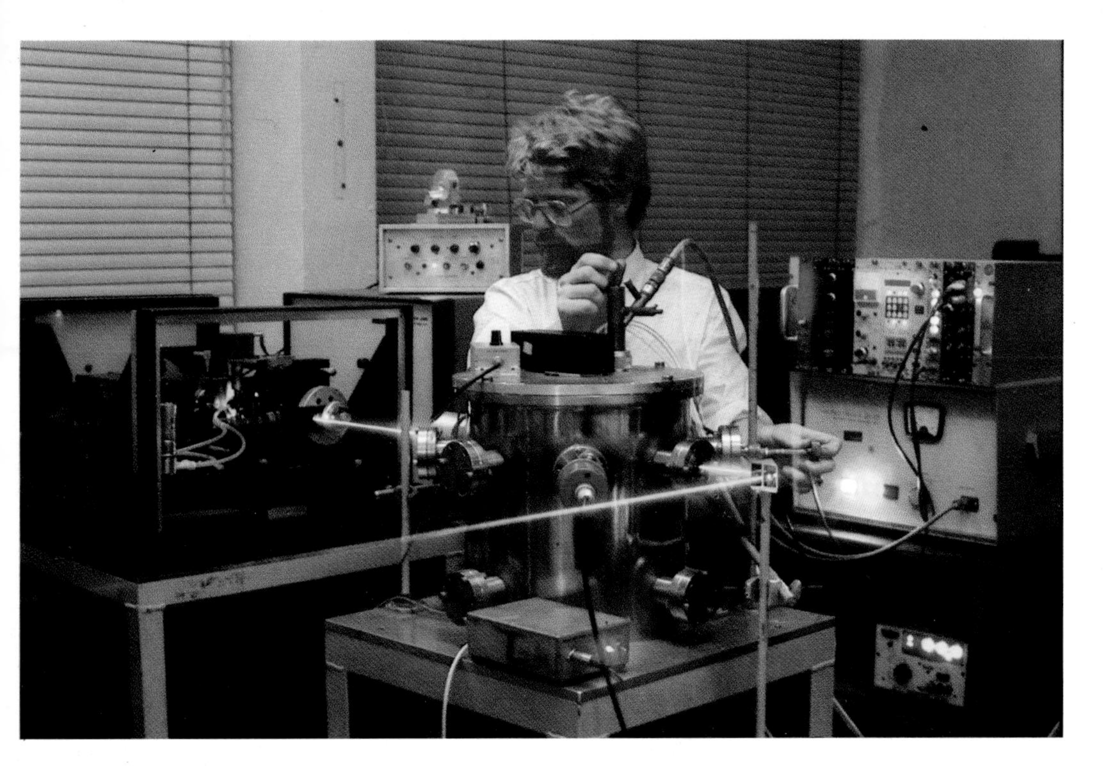

Tunable dye lasers have opened up many possibilities in laser chemistry. They can be used to catalyse reactions—that is, they can be used to excite molecules into states from which they react with other chemicals many times faster than they would normally. A tunable laser can also be used to study the progress of a chemical reaction; it detects the amount of rotational and vibrational energy given off by the newly formed chemicals.

Using laser light in this way is much more efficient than using conventional chemical catalysts.

Already a laser process has been developed in West Germany which has the potential to produce vinyl chloride—a chemical used to make plastics—on a commercial scale. Lasers may also be used to help make ammonia for fertilizers, a process which now requires a great deal of energy. This could be done by using a laser to excite nitrogen to a state in which it reacts much more readily with hydrogen to form ammonia.

Watching Chemical Reactions

Lasers may also be used in a different way to improve scientists' understanding of chemical reactions. Chemical changes often take place very quickly—in a few picoseconds (a few millionths of a millionth of a second). Lasers can now be made to produce pulses that last for less than one picosecond. This means that a chemical reaction can be started by a pulse of light which lasts for less time than the reaction it triggers. So immediately after the pulse of light has passed, scientists can actually watch the reaction take place. This is done by watching the light emitted by the reaction. With longer pulses the laser light would still be around while the reaction took place and the two lots of light could not be distinguished from each other.

This technique has great potential in many complicated chemical and biological processes. In chemistry, reactions that involve colour changes in liquids and solids are particularly suitable for studying in this way. In biology, one area of great interest is the way in which we and all other creatures see light. This involves complex chemicals that respond to light in amazing ways. By watching the response of these chemicals to extremely short pulses of laser light scientists hope to understand these processes and hence reveal more about one of the wonders of the natural world.

High Power Lasers

When the James Bond film *Goldfinger* was first released, high power laser tools of the type used to threaten Bond's life were still only a science fiction dream. But only a few months later the first carbon dioxide laser was produced and this laser could cut through materials. Such lasers have helped to give laser technology its "death ray" image. But in reality, high power lasers have more down-to-earth, though more practical, uses, such as drilling and cutting metals.

A number of different types of lasers can produce high power levels both in continuous operation and in pulses. The most frequently used high power lasers are carbon dioxide lasers that work at a wavelength of 10.6 micrometres—in the mid-infra-red part of the electromagnetic spectrum. These lasers are small enough to be practical, but they can produce tens of kilowatts of continuous output. Another great advantage is that they generate light very efficiently; up to 25 per cent of the input power comes out as laser light. Carbon dioxide lasers can also be used to generate high peak power pulses. However, they are by no means the only lasers that can do this.

Ruby lasers, of the type originally developed by Professor Maiman, can generate high peak powers and have been put to a wide variety of

A high power pulsed ruby laser punching a crater in stainless steel. The energy input to the laser is 10 kilojoules, whereas the output is only 10 joules. But the beam is so finely focused (the diameter of the crater is only half a millimetre) that it generates a temperature of 2,650°C—sufficient to simply vaporize the steel where the beam strikes.

A carbon dioxide laser being used to cut shapes from a sheet of steel. Because no mechanical parts touch the steel, the cutting beam can be manoeuvred in any direction very easily and accurately. The edges of shapes cut in this way are much smoother than those cut by a saw and need very little further treatment.

uses. But the highest powers are produced by lasers that use glass as part of the laser material. One of these in particular, the neodymium-in-glass laser, which works at a wavelength of 1.06 micrometres (in the near-infra-red part of the spectrum), has been the subject of much research. Neodymium-in-glass lasers are being used in huge laser systems that are being developed to study fusion (see page 42). These systems can produce, for a tiny fraction of a second, power levels of millions of megawatts—many times more than the total electrical generating capacity of all the power stations in the world.

Energy from High Power Lasers

In view of the high power levels mentioned above, it may seem surprising that the amount of energy emitted from a high power laser is not actually very large. However, high power levels can be generated, not by the laser itself, but by the way in which the available energy is delivered to the target. In the case of a typical carbon dioxide laser used to do heavy metal cutting or welding, the output level is about 10 kilowatts. This is about the same amount of energy as can be generated by turning on all the hotplates of a domestic electric cooker. But the important thing is that this laser energy can easily be focused down to a spot much smaller than the head of a pin. This concentration of energy is so great that it instantly vaporizes any metal or other material.

In the case of *pulses* of laser light, the amounts of energy are even smaller. Even with the most powerful laser system, if the amount of energy in a single pulse is delivered over a period of a few seconds, it will only heat up a jug of water. But when this pulse of energy is squeezed into about a thousandth millionth of a second, for that instant of time an incredible peak power level is produced.

Cutting, Drilling and Welding

Such immense power concentrations have many varied applications and carbon dioxide lasers are now widely used in industry. Their compact size means that laser machines can be used in factories in the same way as other machine tools. They can cut and drill metals with great precision and very rapidly. In addition, they can handle complicated shapes; a light beam can be moved around much more easily than a mechanical cutter, as it has no particular cutting edge.

Other features of laser cutters make them even more valuable in engineering. Because a laser cut is achieved by either melting or vaporizing the material, it is remarkably smooth and free from rough edges. This is useful when cutting or drilling ordinary metals, and of even greater value when handling harder or more brittle materials. Lasers can cut or drill glass, ceramics or super-hard metals as easily as softer materials. And, interestingly, they are also highly suitable for cutting very soft materials, such as cloth, expanded plastics or paper. A laser beam puts little or no pressure on the material and can be adjusted to cut quickly and cleanly through a great thickness.

Another advantage of using lasers is the fact that the amount of energy in a laser beam can be accurately controlled. This property is of great value not only in cutting but also in drilling tiny holes and precision welding work.

The range of applications is extraordinary. Not only have laser tools been developed to improve the cutting and welding of materials, but

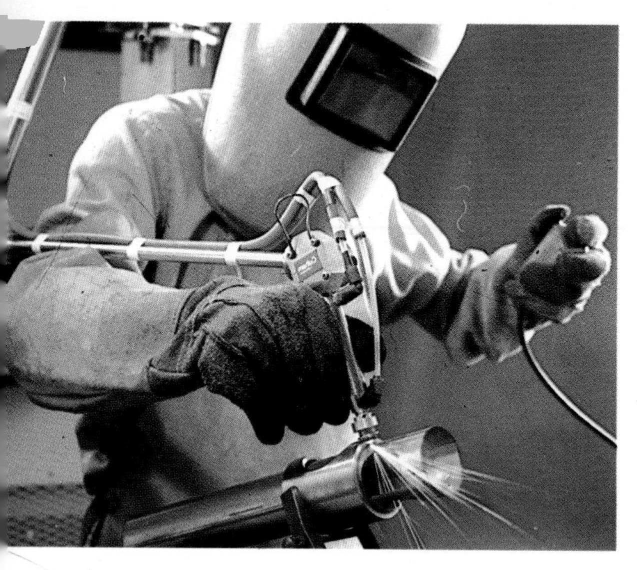

Above: Laser beams can be directed along short lengths of optical fibre with almost no loss of power. Here the beam from a carbon dioxide laser is being fed along a flexible laser beam guide. The operator has complete freedom of movement as he cuts off lengths of pipe.

Above right: A carbon dioxide laser being used to cut canvas. One important advantage of using a laser beam to cut soft fabric is that the heat of the beam automatically seals the cut edges, preventing them from fraying.

also there are lasers to do completely new tasks. The ability of lasers to drill tiny holes is being used by drug manufacturers to make holes in capsules, so that the medicine inside can be released into the body in a more controlled way. To improve electronic equipment, pulsed lasers are being used to solder the connections between components. The exact control of the amount of power delivered to each connection produces a better join. And the laser system can work at high speed—ten times as fast as the best mechanical systems. But the greatest advantage of lasers is in welding. The control of the amount of power gives much better welds. In addition, the fact that laser beams can be directed round corners by using mirrors means that high

Left below: A laser beam can cut away the outer wrapper of a nuclear fuel assembly, but leave the inside unharmed.

Below: Under remote control a laser cuts away part of a nuclear fuel assembly to provide access to the fuel rods inside.

quality welding can be done in otherwise inaccessible places.

The flexibility of laser cutting and welding may have a particular advantage when working in hostile and dangerous conditions. In radioactive areas, for example, everything must be done by remote control and with plenty of protection between the operator and the work. Using lasers to cut up highly radioactive spent nuclear fuel rods could make this difficult work easier.

Treatment of Surfaces

The use of lasers to heat up surfaces rapidly has resulted in a completely new way of changing the properties of materials. When the surface of a crystal is heated to its melting point in a millionth of a second and then allowed to cool very rapidly, the structure of the crystal surface is altered. This process, known as *laser annealing*, has been used to improve the performance of crystals that are used to generate electricity from sunlight in solar cells.

It has been found that laser treatment of such crystals produces much better solar cells. The sudden heating and rapid recrystalization removes many of the defects in the crystal surface. This is very important as it improves the efficiency with which the crystal can convert sunlight into electricity; improvements of 50 per cent have been obtained. What is more, this treatment is inexpensive, as it can be used in the mass-production of solar cells.

The same technique can also be used to make new and better crystals. If a thin layer of special material is put onto a crystal, then a laser can be used to produce a new form of crystal surface from a mixture of the special material and the crystal material immediately beneath it. This process produces new and better crystal surfaces which cannot be made by any other means. In the future such improved crystals may be used to make better micro-electronic devices.

Laser treatment of surfaces also offers exciting possibilities in the control of electroplating. This widely used industrial process uses an electric current to put a thin layer of special metal, such as chromium, silver, nickel or copper, onto some cheaper base metal. It is frequently used in making electronic components. Scientists have found that when small areas of the metal surface are heated during the plating operation, the special metal is deposited in much greater quantities on the heated areas. In future, therefore, laser heating may be used in the electroplating of electronic components to improve their quality.

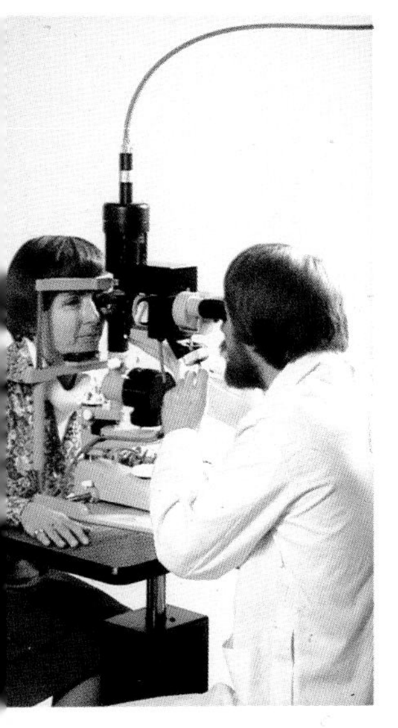

An argon laser being used to treat a disorder of the eye. The laser beam passes through the transparent cornea at the front of the eye, but because the beam is unfocused at this point the cornea is totally unaffected. At the back of the eye, however, the beam is focused to a fine point, allowing the doctor to treat a minute portion of retina (the light sensitive part of the eye).

Lasers in Medicine

One of the earliest uses of high power lasers was in certain areas of surgery and dentistry. The ability to deliver exact amounts of energy with pinpoint accuracy is of particular value in delicate operations on certain parts of the body, such as the eye. The energy can be used to burn away tiny blood vessels that might cause blindness if they spread and to "weld" pieces of detached retina back into place. In dentistry, too, a surgeon can burn away damaged tooth material with pinpoint precision.

These types of operation have become normal surgical practice. But there are now many other ways of using laser in medicine and the list continues to grow. In Austria, lasers have been used successfully to remove "inoperable" cancers of the brain. A laser is used to burn away paper thin layers of the cancer tissue in places where even the tiniest

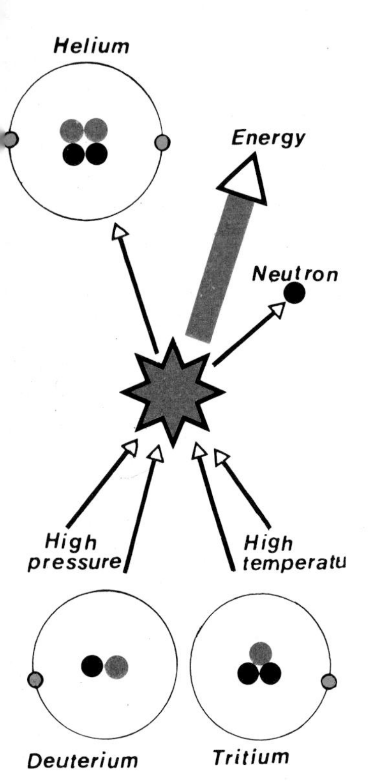

Above: The fusion reaction between deuterium and tritium produces vast amounts of energy.

cut could be fatal. Lasers are also being used to remove birthmarks and tattoos, with considerable success. In future they may even be used to perform acupuncture. Tiny glass fibres could be used to direct light to the desired points on the skin and this would stimulate the nerves in the same way as acupuncture needles do at present.

Laser Fusion

One of the most exciting prospects for high power lasers is that they may one day provide the way of obtaining limitless power from the same process that drives the Sun. This process, which also occurs in a hydrogen bomb, is the result of deuterium and tritium (the two heavy isotopes of hydrogen—see page 33) atoms combining, or fusing, to form helium atoms. This reaction produces vast amounts of light and heat (thermal) energy as the nuclei of the deuterium and tritium atoms fuse together. Hence it is known as *thermonuclear fusion*.

The world's supplies of fossil fuels are being used up rapidly. So there is great interest in the possibility of using thermonuclear fusion as a source of power.

Deuterium is a very common material—there is one gram of deuterium in every 36 litres of sea water—and the potential fusion energy present in a gram of deuterium is equivalent to about 80 tonnes of TNT or 11,300 litres (2,500 gallons) of petrol. Tritium is not available naturally, but it can be produced from the common element lithium. Therefore, in theory, nuclear fusion represents an almost limitless source of power.

However, there is a major problem. Although fusion has been made to occur in hydrogen bombs, it is much more difficult to produce in a controlled form. To get deuterium and tritium to fuse, they must be squeezed together at incredibly high temperatures (about a hundred million degrees centigrade). Scientists are looking at various ways of doing this. One widely researched method involves forming the fusion ingredients into a plasma (a very hot gas in which some or all of the atoms have lost electrons) and squeezing this plasma in a huge magnetic field. A vast amount of work has been done on this process and much progress has been made. But the goal of getting more power out than is put in is still many years away.

Lasers offer a possible short cut to fusion. The intense light from powerful lasers may be able to compress and heat frozen pellets of

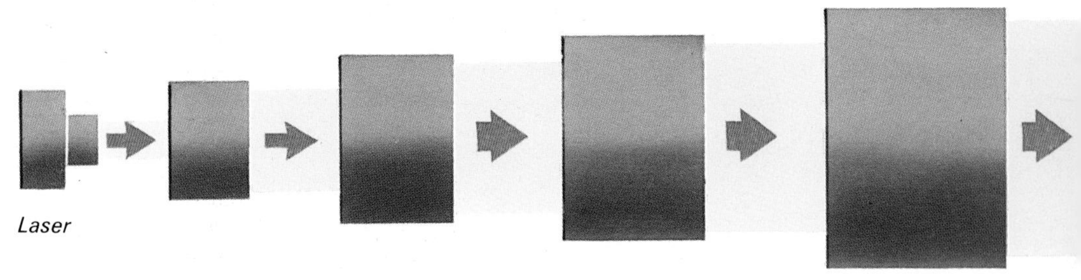

In the proposed laser fusion reactor a number of high energy beams of laser light are focused into a reactor vessel. Each beam is produced by passing the beam of a high power laser through several amplifying lasers—one such laser chain is shown here. When the beam emerges from the last laser in the chain, it is focused into the centre of the reactor vessel. There it strikes a pellet of deuterium and tritium, which has been introduced into the reactor vessel via an injection tube. The effect of the high temperature and high pressure created by the laser beams cause the fusion reaction to occur. Energy is generated in bursts as each pellet is fed into the reactor and burned by the laser beams.

deuterium and tritium to temperatures high enough to produce tiny, controlled thermonuclear explosions. Hundreds of millions of pounds are being spent in a number of major countries to build the massive laser systems needed to do this. Carbon dioxide and neodymium-in-glass lasers are being used in this work and scientists are trying to build an arrangement that can bombard a pellet simultaneously from many angles with ultra-short pulses of light.

However, for fusion to occur, these pulses must be many times more powerful than anything so far produced by a single laser. So to get such extraordinary power levels, it is necessary to use a large number of lasers in series. First in the line is a relatively small laser which can produce a clean, sharp pulse. This pulse then passes through a series of increasingly larger lasers, each of which amplifies the signal. These amplifiers do not have mirrors at either end like normal lasers, but merely work as the pulse passes through them once and once only.

One interesting problem with such a laser system is to prevent it from "back-firing". In the single pass of laser light through the system, not all the excited atoms or molecules are stimulated to emit light. This means that if any of the laser light is reflected backwards from the parts of the lasers, or from the target, it will grow and grow as it travels back along the line. This could result in a huge pulse of energy which would blow up the first laser. To stop this happening, special components are put into the system. These only allow the light to travel in the right direction.

Another great problem is that, as a pellet starts to implode (explode inwards) under the pressure of the massive burst of laser light, it produces a cloud of electrons that reflects some of the light away One

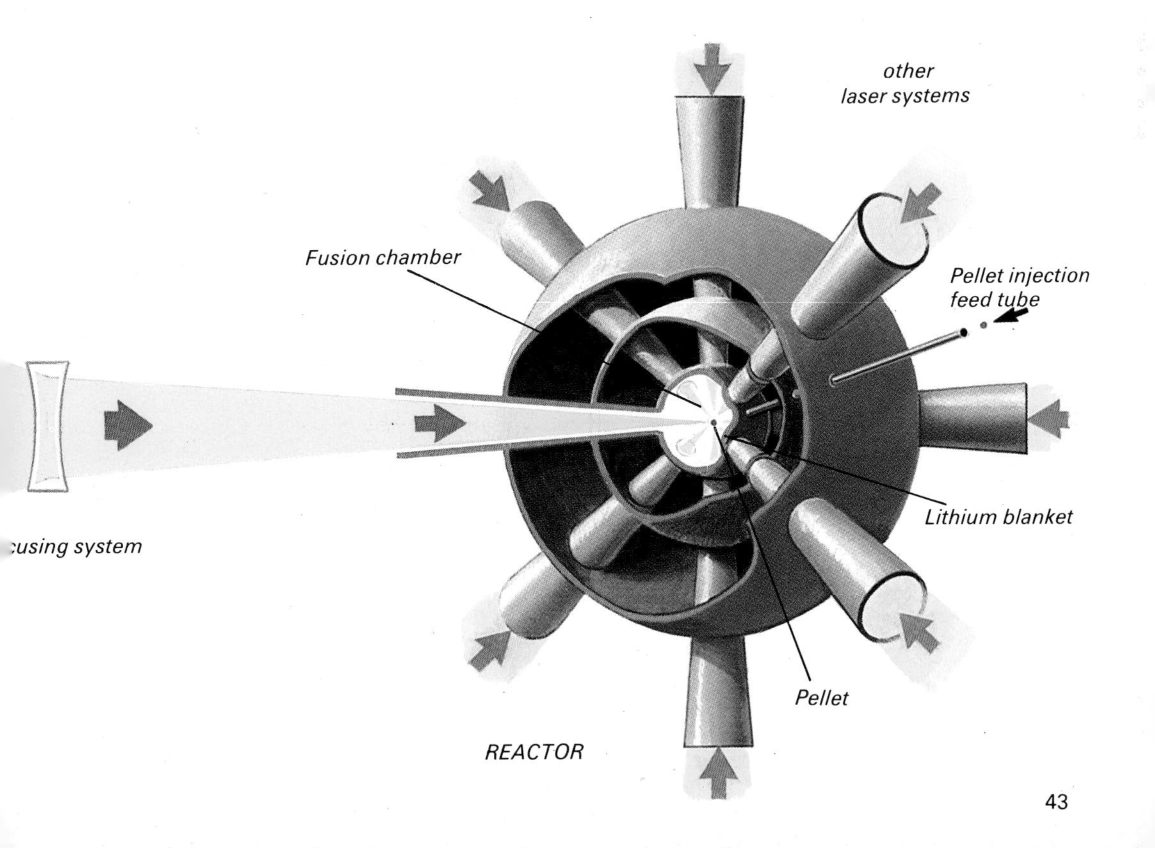

Right: The reactor vessel of the Shiva laser.

Far right: Inside a reactor vessel. The lenses focus the laser beam onto a point just in front of the injection tube (top right). Sensors placed at intervals around the chamber monitor the reaction.

way of overcoming this problem may be to use powerful lasers with the shortest possible wavelengths. The neodymium-in-glass laser already has a relatively short wavelength (1.06 micrometres). And, in fact, scientists are investigating ways of making this laser produce even shorter wavelengths by harmonic generation (see page 17). It is hoped that a wavelength of 0.35 micrometres (one third of the laser's natural wavelength) can be produced with little loss of power and that this will be a major step towards achieving laser fusion.

In theory, if enough light can be delivered to the target in sufficiently short time, the light will compress the pellet and heat it to such extraordinary temperatures that nuclear fusion reaction will take place. However, to achieve a net output of energy from the fusion reaction, larger lasers must be developed. And scientists must learn more about the fusion process itself, so that they can use the laser light as efficiently as possible.

Below: A 1000-joule neodymium-in-glass laser chain. Inside each amplifying laser is an oval disc made of special glass containing atoms of the element neodymium. Around this are powerful flash lamps, which provide the excitation energy for amplifying the laser beam.

Laser Weapons

Almost from the time the first laser was built, the possibility of "death ray" laser weapons has been widely publicized. But the image of a laser beam as a "death ray" is not a true picture. Although laser power levels have been developed to perform heavy cutting and welding tasks, their potential as weapons has yet to be clearly shown. In fact, there is much argument about whether lasers will ever make useful weapons.

But, while the argument continues, the United States and the Soviet Union are spending huge sums of money on trying to develop laser weapons systems. The reason why lasers are so attractive as potential weapons is that they offer the possibility of delivering high powers over great distances at the speed of light—almost instantaneously. This makes them of particular interest in providing defence against missile attack; existing defence systems are much slower.

Carbon dioxide lasers are the most commonly used lasers in weapons studies. A great deal of work has gone into improving the performance of these lasers. The aim is to make them more powerful and, at the same time, take full advantage of their high efficiency.

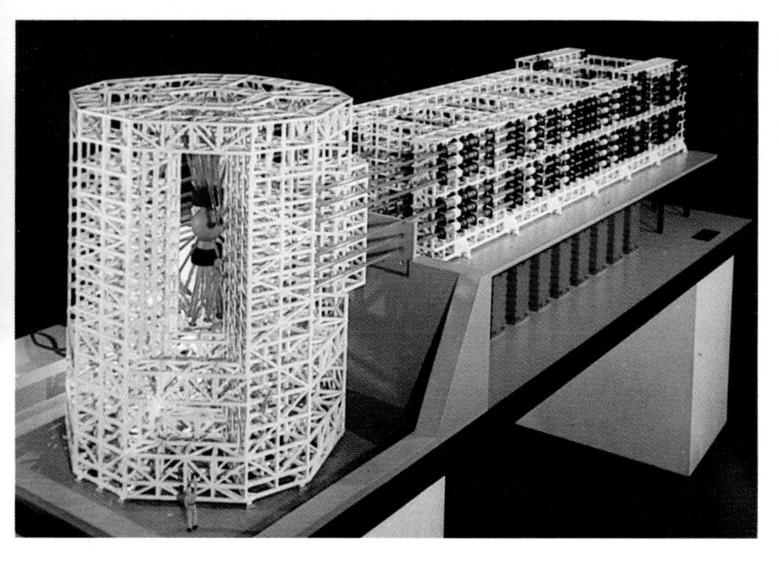

Left: A model of the 25 trillion watt Shiva laser that is being used at the Lawrence Livermore Laboratory in California, for fusion research. It has 20 powerful laser chains.

Nevertheless, to produce the power levels needed to knock out missiles quickly and reliably still requires a very large carbon dioxide laser indeed.

One way of producing a more compact system with a high power level may be to use a chemical laser in which a large amount of energy is available in chemicals that react explosively to produce a high level of excitation energy (see page 36). For example, in a hydrogen fluoride laser the reaction between the gases hydrogen and fluorine is particularly explosive. A lot of work, much of it secret, has been done on this laser to produce very high power levels. Other kinds of laser are also being tried out.

There are several other problems involved in developing laser weapons. The most obvious one is the fact that the atmosphere absorbs many wavelengths. Here the carbon dioxide laser has a marked advantage; the atmosphere is almost transparent to the wavelength it produces. But all laser wavelengths are strongly absorbed by clouds, so it is unlikely that lasers will ever be used in ground-based weapons systems. If, however, they can be flown above the clouds in aircraft or satellites, they may have a role to play. This is why so much time is being spent trying to make more compact and efficient systems.

Powerful laser systems are being tested in large aircraft and it is believed that the Russians have devised "killer" satellites that use lasers to knock out other satellites. Clearly, a great deal of work is going on. But major advances will have to be made before viable weapons systems can be built. In defence weapons systems, for example, the scale of operations needed to set up an anti-ballistic missile system will be enormous. Defence experts in the United States have suggested using a system of 25 satellites, each armed with a five megawatt chemical laser that would be capable of destroying up to 1,000 intercontinental ballistic missiles all launched at the same time.

All this may sound like science fiction. But the amount of work being put into solving the technical problems may well result in some practical form of weapon being developed. Then the images of films like *Star Wars* may not seem so far-fetched.

Above: In the future there could be laser defence systems in space. A number of powerful lasers mounted on satellites could knock out hundreds of nuclear missiles in a very short time.

Guided Missiles

An aspect of laser weapons systems that is far less fanciful is the use of lasers to guide missiles. Already armed forces around the world are using a wide variety of laser systems to enable missiles to home in on their intended targets. The principle of this guidance is quite simple. The target is illuminated with laser light and the missile has a detector that senses this light. The detector feeds information electronically to the control systems of the missile, which then homes in on the target. Depending on the circumstances, the target may be illuminated by a laser in the aircraft launching the missile or a laser operated by ground forces.

The great advantage of laser guided missiles is their increased success rate. Ordinary missiles with less complicated guidance systems rarely hit their intended targets and so a large number have to be used to ensure that at least one strikes home. With guided missiles the number needed to destroy the target is greatly reduced.

Below: A technician using an optical fibre flashlight to check a GLLD (Ground Laser Locator Designator)—a device used to mark targets for laser-guided missiles.

Left and below: Laser levitation. A tiny particle can be suspended in a laser beam due to the pressure of light around it. The tiny star-like glass particle in the photograph is being held in mid-air by an invisible laser beam. This particle is only as wide as a human hair.

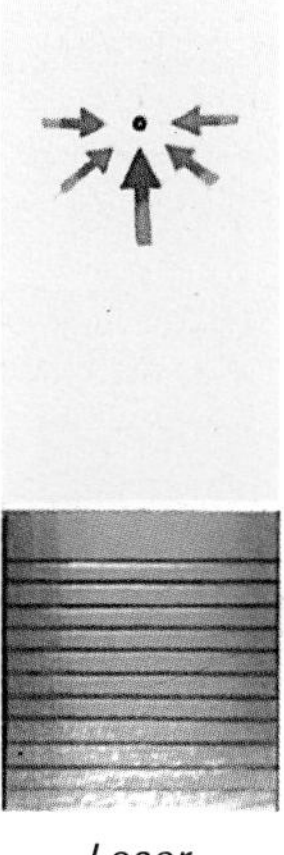

Laser Levitation

Yet another intriguing aspect of high intensity laser light is its ability to levitate small objects. When light strikes a reflecting surface, it exerts a tiny pressure. In the case of sunlight this pressure is too small to have any appreciable effect. But with much more intense light the pressure is much greater and can be adjusted to balance the force of gravity. This means that a laser beam can be used to suspend small particles in mid-air.

In fact, this feat does not require exceptionally high-power lasers. Continuous power levels of only a few watts are needed to suspend particles of about a tenth of a millimetre in diameter. And only a few microwatts are needed for particles of half a micrometre in diameter. Levitation is due, not so much to the power of the beam, but to the uniformity of the brilliant light. This traps a particle by holding it with even pressure on all sides. The result is that the beam acts rather like a fountain in a fairground shooting booth on which a ping-pong ball is

Below: An operator marks the target by illuminating it with a spot of laser light. A laser-guided missile has a special sensor that detects the laser light reflected from the target and homes in on the spot.

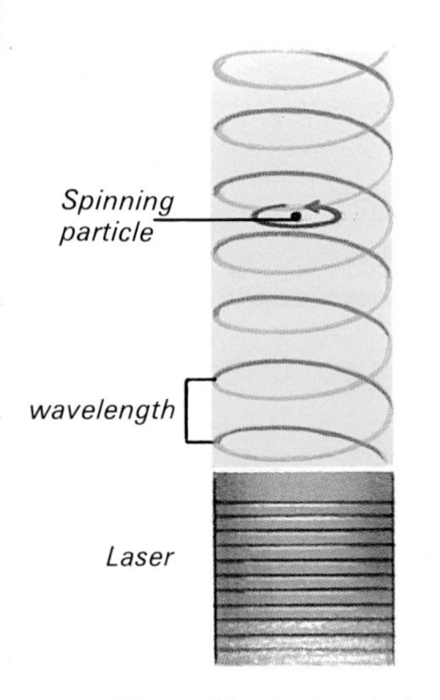

Above: Circularly polarized light can be thought of as having a spiral wave pattern. If a particle is suspended in a beam of this kind of light, it tends to spin as the spiral of waves moves away from the laser.

Below: A beam of atoms tends to spread as it moves away from its source (left). But if it is surrounded by a focused beam of laser light, the pressure of the light pushes the atoms inwards so that they are collected over a smaller area on the plate (right).

suspended. However, even though the power level is not terribly high, the strength of the beam can still melt or vaporize opaque particles. So such levitation can only be performed with transparent particles.

Levitation can be used to study the way in which light is scattered by small particles. The process of light scattering by dust and other small particles in the atmosphere is currently a subject of intense scientific investigation. Among other things it affects visibility in fog and mist and the way in which the Sun's energy is absorbed in the atmosphere. Hence it also affects how dust and clouds influence the Earth's climate. Before the advent of lasers there was no way of suspending a tiny particle absolutely still in mid-air without any mechanical support. But now the precise scattering patterns of a single particle can be measured—with an increase in accuracy of betweeen a hundred and a thousand times.

The same technique can also be used to levitate liquid droplets. This is particularly interesting as it can be used to study the way in which droplets either grow or evaporate, depending on the temperature and the amount of water vapour in the surrounding atmosphere. It can even be used to watch how droplets join together when they bump into one another. These studies have many applications. The stability of droplets is of great interest to scientists trying to learn more about such things as the growth of clouds and the properties of air and liquid fuel mixtures.

The use of light pressure has even more interesting possibilities. Laser levitation can be used to suspend a tiny sphere in what is known as *circularly polarized light*, which can be produced by certain laser systems. The wave train of this light can be visualized as a spiral pattern. The distance between successive coils of the spiral is equal to the wavelength of the light. When a particle is suspended in such light, the spiral wavetrain (which is, of course, moving in the direction of its axis at the speed of light) causes it to start rotating. But as the particle rotates faster and faster, it is subjected to enormous strains. And, long before it reaches a spin speed equal to the speed of light, it disintegrates. It has been suggested that this process could be occurring in interstellar space, where it has the effect of limiting the size of the grains of dust in these regions. Back on Earth, the process offers scientists an interesting way of studying the strength of materials.

Another way of using the pressure of light is to guide a beam of atoms or molecules. If the laser light has a frequency close to that which is absorbed by the atoms or molecules, then the laser beam can be used to push them around. This effect can be used to focus a beam of atoms or molecules and enable them to be collected more efficiently. The same effect may also be used to separate isotopes (see page 33).

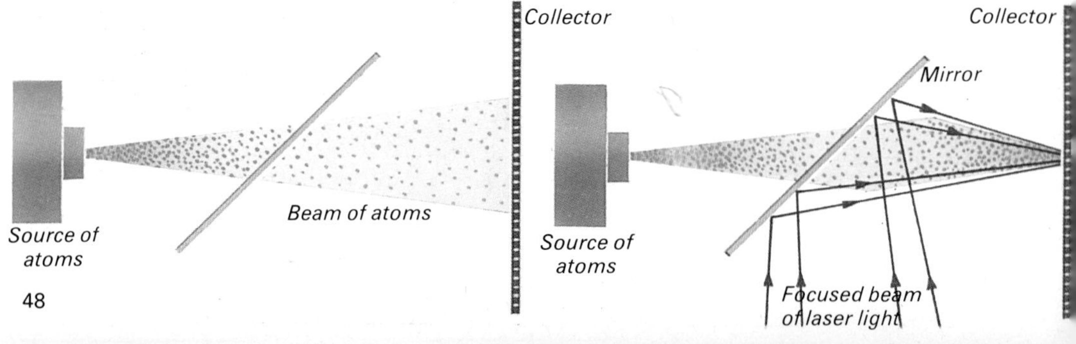

Laser Information Systems

The technology of using light to send, receive and process information is rapidly expanding. And lasers are playing an increasingly important part in this technology. They are being used to read and record information. And they have a crucial role in the sending of information.

Optical Communications

The principle of using light to send information is very simple. If you use a mirror to direct flashes of light at someone to attract their attention and send a simple message, you are using the simplest form of optical communications. The Aldis lamp, which has for many years been used to send signals between ships at sea, works on the same principle. Today, lasers are being used in place of these more simple devices.

Using lasers to transmit information has three important advantages. First, a single laser beam can carry huge quantities of information. Second, the technique of modern electronics and the controlability of certain lasers make it possible to put information onto a laser beam at high speed. Third, the fact that a laser produces a very narrow beam means that it can be directed to only a single receiver.

Lasers are not alone in the field of optical communications. They face a certain amount of competition from some sources of ordinary light. The most important of these are electronic devices called

A laser beam traffic control system in Atlanta, Georgia. The system uses low-intensity laser beams to transmit messages from a central computer to the junction controllers, which operate the traffic signals. The computer sends out instructions as it receives information about traffic flow from detectors in the roads. The use of lasers instead of overhead wires or underground cables reduces cost and installation time. The system works in heavy rain, snow, lightning and high winds. On a clear day the beams are capable of transmitting messages for over 15 kilometres (10 miles). Even in thick fog they will still work over a distance of more than 600 metres.

light-emitting diodes, which are widely used in such things as digital displays for watches, calculators and computers. Light-emitting diodes can largely match the carrying capacity, speed and controlability of lasers. In addition, they are very small and can be made very cheaply, so they offer lasers severe competiton.

Lasers, however, are capable of doing much more than light-emitting diodes. Since they generate "cleaner" light at higher power levels, the light can be transmitted over great distances without having to be amplified along the way. This is important when light is sent through long glass fibres (see page 52). Also, they can handle information much more quickly. So it may be that lasers will be used for one type of work—long distance transmission of large amounts of information—while diodes will be used for shorter, simpler tasks.

Competition between lasers and light-emitting diodes may have another advantage. The fact that, in certain areas of communication technology, either lasers or diodes can be used, will increase the rate of progress in these areas. Manufacturers of both lasers and diodes will be spurred on to developing better and better devices.

Where a laser has no competition is in its narrow beamwidth. A laser signal can be directed with high accuracy and efficiency at the receiver. Only low powers are needed to send the signal over long distances and it is almost impossible to intercept messages without the sender and receiver knowing. By way of contrast, radio communications are broadcast in all directions and so can be picked up by anyone who has a suitably tuned receiver.

However, sending information along a laser beam has its limitations. Communication can only take place across an open space— any obstruction, even clouds or fog, prevents the light from reaching the receiver. As a result laser beam communication is principally of military use. Using small, portable transmitters military personnel can communicate with each other over short distances without their messages being intercepted by the enemy. But simple systems may also be used in big cities to provide telephone links between office blocks for companies who want a cheap, secure means of sending confidential information.

Above: A powerful beam from a continuous-wave argon laser carries a message to a satellite orbiting the Earth. This laser has been used with Explorer satellites and others. It has also been used to measure the distance from Earth to the Moon, using the mirror left by the Apollo 11 astronauts.

Below: Some optical fibres are made from glass preforms. The preform is heated until soft and then drawn out into a thin fibre. Here, red laser light is being used to test the optical quality of a preform.

Messages Through Glass Fibres

The problems of using laser beams to carry information largely disappear when, instead of beams being sent through open air, they are sent along glass fibres. These are hair-thin fibres made from special kinds of glass. Light can travel along such fibres with very little loss of intensity. This offers the possibility of using glass fibres instead of expensive copper cables to carry telecommunications.

Glass has only recently been used to carry messages. The reason for this is that ordinary glass is not very clear. A pane of glass may look transparent, but when seen from the edge it has a greenish-blue colour. This colouring is due to the fact that impurities in the glass absorb quite a lot of the light.

To make low-loss optical fibres these impurities must be removed. The laser provided an impetus for scientists to examine ways of doing this. The technology of making pure glass progressed rapidly and scientists can now produce glass fibres which can transmit light signals at certain wavelengths up to 50 kilometres before they need to be boosted. These losses are much lower than those in copper co-axial cables, which are currently used for transmitting information at high

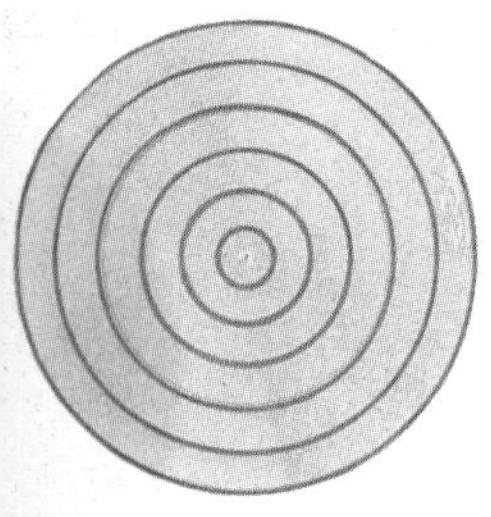

Cross-section

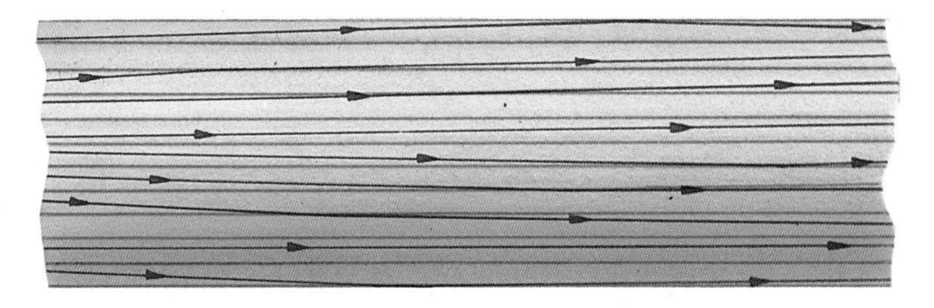

rates. The losses in these copper cables are such that a booster—a device known as a repeater—is needed about every two kilometres.

To achieve such spectacular advances, it has been necessary not only to remove all the impurities from the glass but also to manufacture the fibres from several different kinds of glass. Each fibre is made up of layers. From the centre of the fibre outwards each layer has a greater refractive index (the property of a material that controls the refraction, or bending, of light that enters the material). This means that, as light travels nearer the surface of the fibre, each layer bends it more nearly parallel to the surface. Any light that does reach the surface does so at such a shallow angle that it is reflected back into the fibre. As a result all the light is guided along the fibre and none leaks out.

The process of making such fibres requires complicated manufacturing methods, but technologists have devised ways of drawing out carefully constructed combinations of special glasses into hair-thin fibres many kilometres long. Such fibres work best with light in the near-infra-red part of the spectrum. Using wavelengths of between 1.4 and 1.6 micrometres, less than half the light is lost over several kilometres.

Other components needed to make optical communications systems are also now available. Scientists have developed tiny light sources, connectors to get the light into and out of fibres without

Above: Sections through an optical fibre. As rays of light travel down the fibre (right), they are bent inwards by each layer. Any that reach the surface are reflected back inside.

Below left. In this technique of fibre-making each fibre is coated with silicon to prevent its surface from cracking. The technician uses a laser to ensure that the coating of silicon is spread evenly all over the fibre.

Below: A technician directs a krypton laser beam along a length of fibre to measure the loss of light.

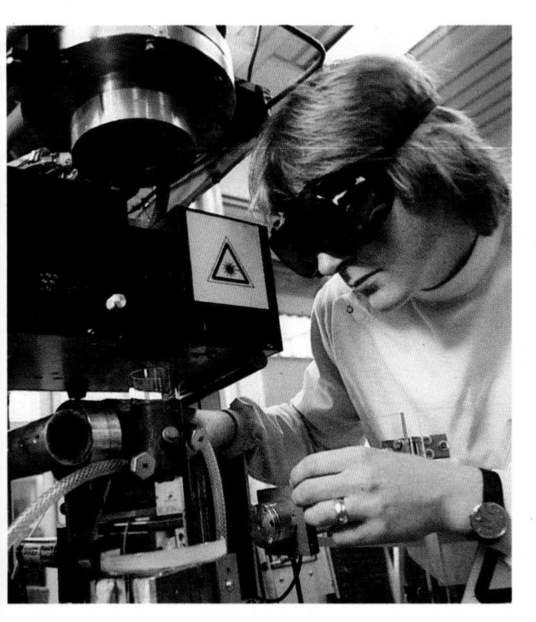

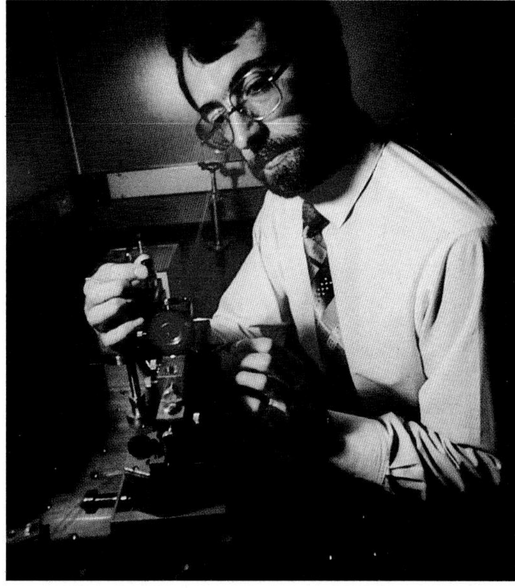

losing too much, and simple amplifier systems to increase the signal along its journey. In many countries demonstration projects are already being set up to test the advantages of optical communications. In one area of Britain some houses are being equipped with cable television in which the television signals are transmitted by laser beams in cables containing many glass fibres. The pace of development is such that in a few years optical systems may well take over in virtually all new telecommunications systems.

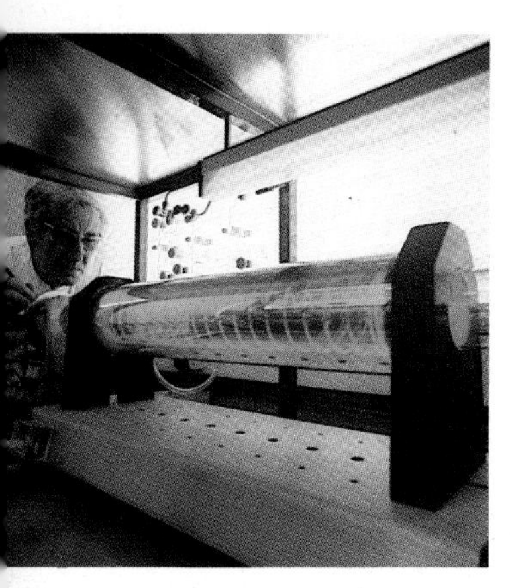

Above: Laser communications systems need tiny lasers. Here, tiny semiconductor lasers are being made inside a gold-lined furnace at a temperature of 1,000°C. Each laser consists of a fragment of laser material surrounded by five or six layers of semiconductor material.

Using Optical Communications

Optical communications are all the same in many ways. The signal to be transmitted is first put into an electrical form, which is then used to control the output of a tiny crystal light source—which may or may not be a laser. The light signal is then fed into a glass fibre, which carries it to its destination. If the signal has a very long way to go, it may have to be amplified on the way. This is done by putting one or more repeaters in the line. At the far end the light signal is converted back into an electrical pattern, which can then be used to create a form of signal that can be understood by the person receiving the information

In such a system a laser has several advantages over an ordinary light source. But possibly the greatest advantage is that a laser can be used to provide a complete, self-contained information and power system. Scientists have developed communications systems in which a laser is used not only to transmit information but also to give enough power to operate all the components in the system. Where power is needed, special devices convert some of the light into electricity, which powers the various components. This has the great advantage that there are no power sources other than the light fed into the optical fibre, which makes the system cheaper and less likely to break down.

There are a number of reasons for developing optical communications. First and foremost, they can provide a much greater information-carrying capacity than existing systems. For example, the capacity of the world's telephone networks many of which are now overloaded, could be greatly increased by replacing conventional cables with optical cables. Cables containing glass fibres are much more compact; the individual fibres are very thin and they can be tightly packed without the signals in one fibre interfering with those in adjacent fibres—a great problem in high-frequency electrical communications. In addition, a laser beam in a glass fibre can, in theory, carry much more information than an electric current in a copper cable. This theory is based on the fact that a laser beam has a much higher frequency. The amazing carrying capacity of optical fibres may never be fully exploited in the world's telephone systems, but it is interesting to consider that a single laser could probably allow half the people in the world to talk to the other half.

A greatly increased telecommunications capacity is likely to be needed for several reasons. Apart from the fact that more and more people want telephones, there is a growing demand for some sophisticated home and business communications systems. These

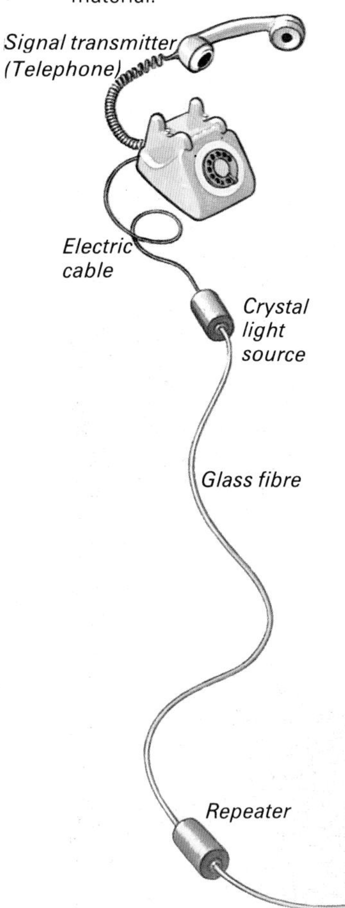

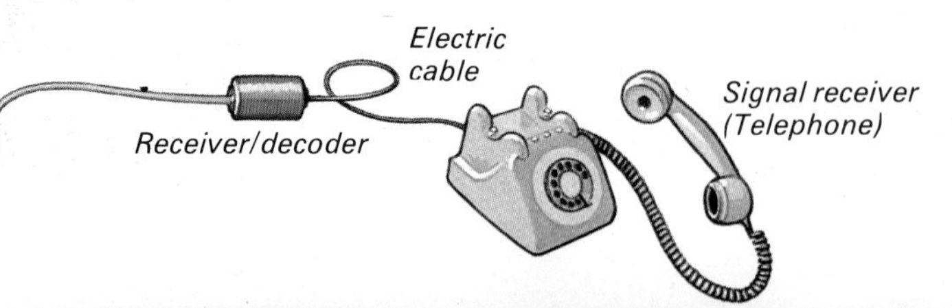

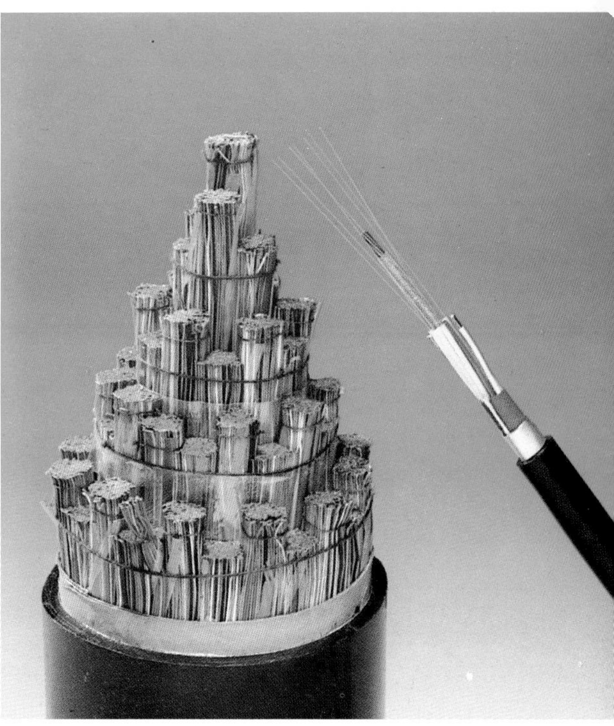

Above right: An optical fibre cable compared with a conventional electric cable. The electric cable is 8.25 centimetres in diameter and can carry 6,000 conversations using 12,000 wires (6,000 pairs). The optical cable is only 1.27 centimetres in diameter and can carry 7,680 conversations in just eight fibres—each pair of fibres carries 1,920 conversations.

Above: A technician operates a fibre jointing machine. The joined fibres are wound onto drums ready for use in optical cables.

Left: The basic features of a laser communications system.

include cable television, video-telephones and machines for transmitting copies of documents and printed pictures. As the expected demand for such equipment grows, so will the pressure for more telecommunications capacity. In the future many business meetings may be conducted more cheaply, quickly and efficiently without the people involved moving from their own offices. Video-telephones will allow everybody taking part to see each other, and copies of documents, including hand-written notes, photographs and drawings will be sent via the telephone lines. Already conferences have been held linking up people in 20 widely separated parts of the world.

The capacity for sending information from one place to another will also have to increase as more and more people and organizations use linked computers. Already large computers in different countries are linked by conventional cable systems to take advantage of their combined capacities. This uses the computers more efficiently and shares the load of storing information. Banks and other financial institutions are conducting more and more business electronically between computers. And in the future credit card transactions will probably be almost entirely handled by linked computers. Such computer networks will have to handle vast flows of information quickly and efficiently and optical communications systems will offer the best way of doing this.

Increasing the capacity for sending information is not the only reason for developing optical communications systems. In fact, they may come into use simply because, compared with copper, glass is cheap, plentiful and light.

So once glass fibres can be mass-produced, they will be used in preference to copper wires. And, for certain applications the

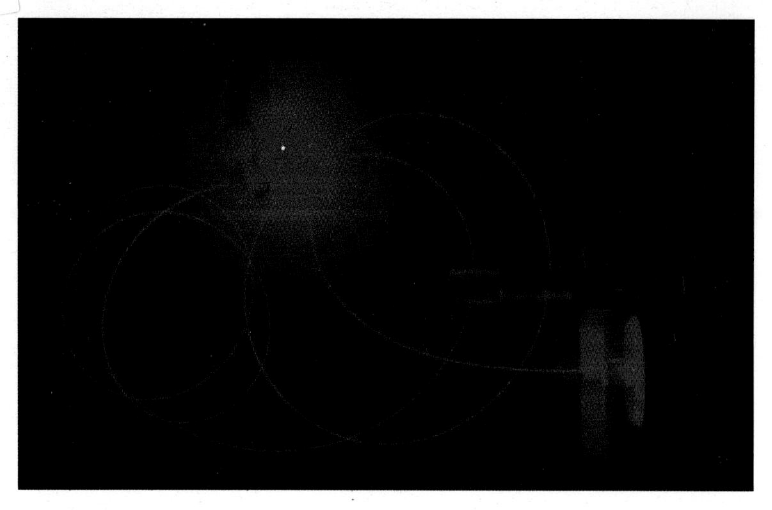

Left: A low-loss optical fibre illuminated with red light.

reduction of weight will be a great attraction. In modern large aircraft there are many miles of copper wiring that could be replaced by much lighter optical fibres.

Optical Fibres in Industry

One further way of using optical fibres and lasers is to make measurements of hazardous industrial chemicals. There are a number of industrial chemical processes that need to be checked regularly to make sure that they are proceeding in the right way. Often this involves looking at corrosive, poisonous or radioactive chemicals and this cannot easily be done in the factory with sensitive, expensive equipment that might be damaged. So samples are taken regularly and taken away to a laboratory for analysis. Even then an accident with the chemicals may damage the equipment.

Now, however, a new method has been developed that allows a laser spectrometer (see page 30) to "see" right into a factory chemical process by means of optical fibres. The spectrometer is kept in a clean, safe place, which can be more than a kilometre away from the factory, and it never comes into contact with the chemicals. Light is fed through a fibre to a convenient point from which the chemical process can be observed. The light that is given off by the chemicals that are present is collected and fed back to the spectrometer. The information recorded by the instrument is used to check that the process is going according to plan.

This equipment combines the ability of a laser to detect the presence of particular molecules by their absorption spectra and the ability of a glass fibre to carry considerable amounts of light over a long distance. Moreover, the cheapness of optical fibres means that if they are damaged in the factory they can be replaced at a much lower cost than conventional measuring equipment.

Writing and Reading with Lasers

Lasers can be used to do more than just send information. They can record, or write, information and their ability to read coded signals has a variety of practical uses. In modern information systems lasers can be involved at many stages.

Specially designed lasers are used to record information that can be stored and retrieved later. The simplest way of doing this is to record the information photographically. Lasers can be used to make permanent photographic records that cannot be erased accidentally.

Lasers systems also offer another attractive way of producing permanent records. Information can be stored in the form of binary code (the two-digit code used by computers) by using a high power laser to burn tiny holes in a metallic film laid on a glass plate. Because lasers can make very small holes at high speed, huge amounts of information can be recorded in much less space than with other systems. Not only is this a very efficient way of handling such information, but also it is virtually impossible to damage the record in the normal course of operations.

Lasers are unlikely to compete directly with normal computer memories. The magnetic tapes and discs used in such memories already provide flexible and fast ways of handling information. But these systems are vulnerable to accidental damage or deterioration and hence are not ideal for permanent records of such things as financial information, personal records and historical knowledge.

Reading by laser is the reverse of the writing operation. The presence or absence of binary digits can be detected by scanning a suitably recorded memory store with a low power laser. Lasers can also be used to read other forms of recorded information. They are commonly used in supermarkets to read the bar code that is stamped on the packaging of many goods. These codes give precise details of the type of goods being sold and this information is fed back to a central computer, which keeps a record of all sales.

Lasers are also being used to read television video discs. In a laser video disc system, the picture and sound information is recorded on the disc as a series of reflecting optical pits. These are arranged in a spiral pattern similar to the spiral groove on a normal gramophone record. Inside the video disc player a laser reads the information on the disc and this is converted electronically into a television signal, which is fed into the aerial socket of the television set. Scientists are now trying to develop video discs for recording in the home.

At a computer-controlled supermarket checkout (top) a laser scanner is used to identify the items and their price. Each item is pulled across the scanner, which is set into the checkout stand. The scanner automatically reads the bar code (bottom) on each item and the information is passed to the checkout terminal. Several terminals are supervized by a central computer.

In an optical video disc player (left) the surface of the disc contains a spiral pattern of reflecting optical pits. A laser beam is directed at the surface of the disc and the pits reflect back a stream of varying light. This is converted electronically into a television signal.

Holography—The New Photography

Above: A scientist making a hologram of some pixie-like figures. The hologram is recorded on a photographic plate.

Among the many exciting applications of lasers is a totally new form of photography, known as *holography*. Normal photographs are two-dimensional pictures. But holography can create three-dimensional images that, until you try and touch them, appear as real as solid objects.

Laser light can be used to produce such images because it is coherent—all the light waves are of the same wavelength and are in phase. An ordinary photograph is made by recording, in two dimensions, the direction of each part of the object from the photographic film or plate and the amount of light reflected by each part. In holography this information is also recorded. But, in addition, the coherent light can record information about the distance (the third dimension) of each part of the object from the photographic plate.

This is achieved by producing two identical beams of laser light, either by splitting a single laser beam or by using two lasers. One beam, known as the *reference beam*, is directed straight at the photographic plate. The other beam lights up the object and light from the object is reflected onto the plate, where it is combined with the reference beam.

But different parts of the three-dimensional object are at different distances from the plate. So when the light waves from these parts combine with the waves of the reference beam they are out of phase by varying amounts. The result is that the photographic plate records an interference pattern that contains information about the distance as

Below: To record a hologram a laser beam is split into two. One beam illuminates the object, while the other beam follows a separate path to the photographic plate. Here, the two beams combine and an interference pattern is recorded on the plate.

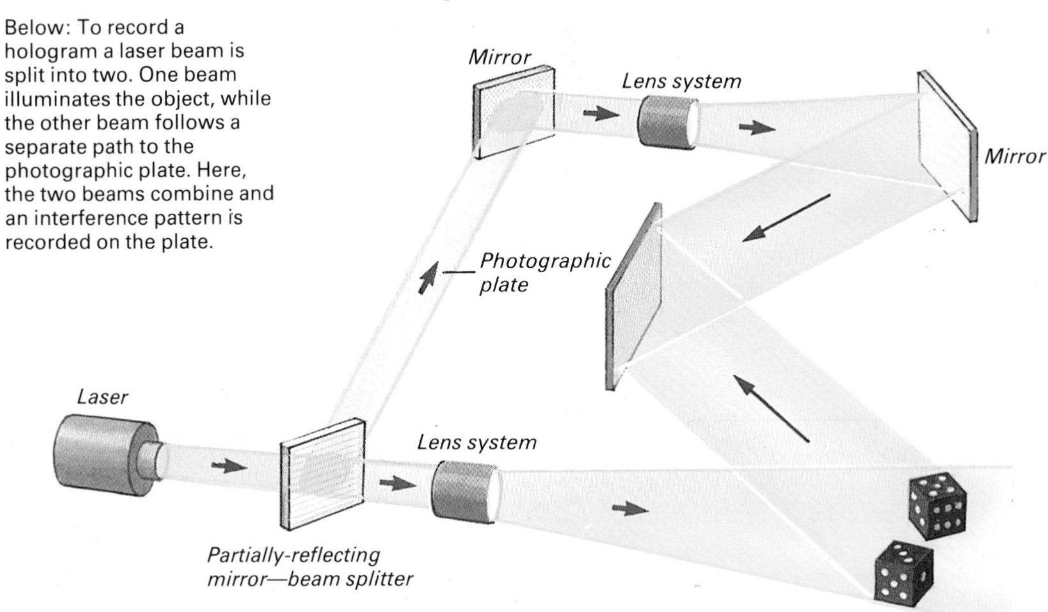

To create a holographic image the hologram is illuminated with laser light. An observer on the other side of the plate sees exactly the same wave pattern as was used to record the hologram and therefore sees an image of the object. But if the observer moves to one side, the light pattern changes slightly. Thus the image also changes and it seems to the observer that he is looking at the same three-dimensional image from a slightly different angle. Here, two observers are viewing the hologram. The images they see (known as virtual images, because the light rays do not actually come from them) are different views of the object and they appear to have depth as well as width and height.

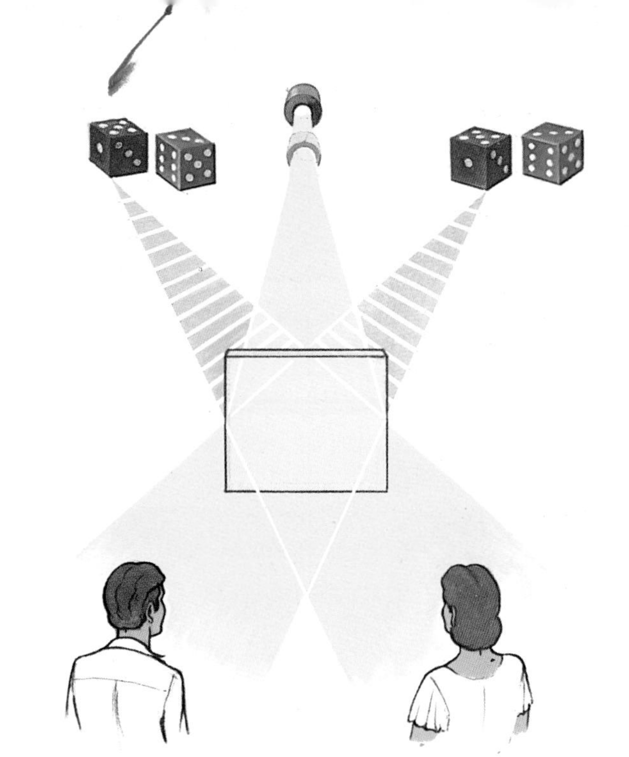

well as the direction of and amount of light from each part of the object—in the same way as the interference signal in an interferometer provides information about the distance of a movable mirror (see page 20). The recorded information is known as a *hologram*.

However, the hologram itself does not look like a picture. The interference pattern has no identifiable information and it looks like a blank, matt surface when viewed in ordinary light. But when it is illuminated with laser light of the same wavelength as the light used to record the information, an observer sees a three-dimensional image. This is produced because the hologram alters the coherent light waves

A holographic image. The three-dimensional image is produced by illuminating the hologram with laser light of the same wavelength as the light used to record the hologram. Here, blue laser light has been used. As a result the image appears in only this colour. Eventually, however, scientists hope to develop a way of creating a holographic image that faithfully reproduces all the colours of the original object.

Right: Several holograms (represented here by coloured sheets) can be recorded in differently orientated planes in certain crystals.

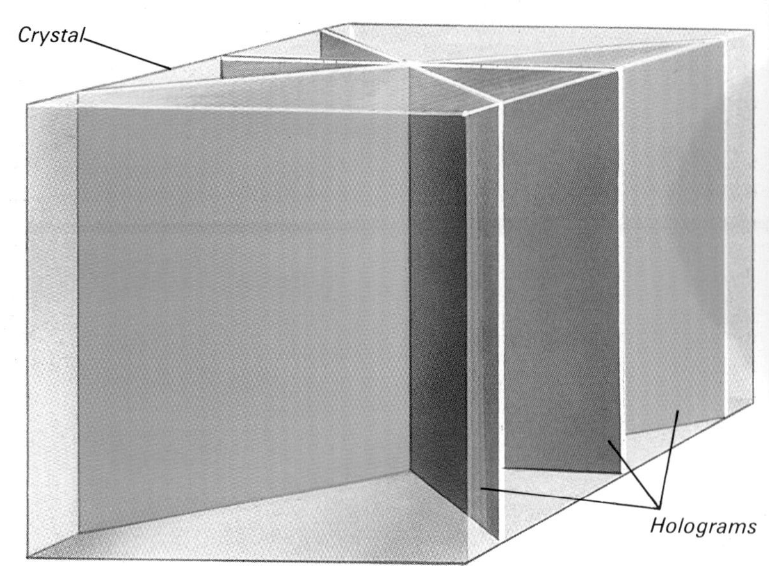

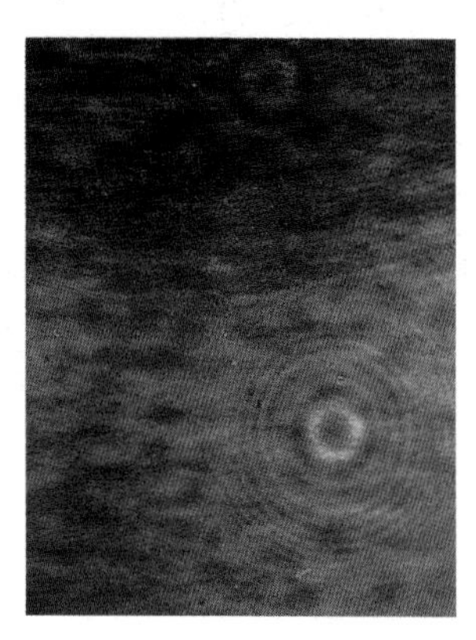

Above: A close-up of a holographic plate, showing the interference pattern. Only when it is illuminated with laser light identical to the light used to record the pattern will the observer see a three-dimensional image.

Below: Holography is useful for storing information in a small space. Here, several holograms (the red areas) have been recorded on a piece of photographic film. The information in any hologram can be retrieved at any time simply by illuminating it with laser light.

which pass through it to produce exactly the same wave pattern as the one that came from the object. This means that when you view a holographic image from a number of different angles it is just as though you were looking at the original object lit up by laser light. So you can see the parts of image appear to move in relation to each other, as if you were looking at a solid, three-dimensional object.

An example of a holographic image is shown in the photograph on the previous page. However, it must be emphasized that it is impossible to show such an image properly in two-dimensional photographs. You must see a real holographic image to appreciate its remarkable properties.

The quality of holographic images has now been perfected to such a high degree that they are often used as the highlight of laser exhibitions. Using two laser beams that cross over each other—one carries the hologram information, the other is an unaltered reference beam—an image can be projected in mid-air. It is almost impossible to believe that there is no object there and at some exhibitions people are invited to reach into the image and prove that it cannot be touched.

A hologram has another remarkable property. Any part on the hologram contains all the information about the scene—the information is therefore spread uniformly throughout the hologram. As a result it is possible to cut a hologram into pieces and yet still be able to reconstruct the scene from a single piece, although the quality of the image decreases as smaller and smaller pieces are used. This property is especially valuable when holography is used to record important information. Even if the hologram is damaged, it is likely that enough of it will remain for the information to be retrieved.

The use of holography to store information is, in fact, an area of great interest. A great deal of work is being done in developing holographic techniques for storing information in computers. One way of doing this is to record the information in special crystals instead of photographic plates. Scientists have found that it is possible to record a number of holograms in a single crystal by recording them on differently orientated crystal planes. In this way the entire volume of the crystal can be filled with information. And the information takes up far less space than it does in existing computer memories.

Holograms may also be used to store confidential information. If the information is recorded partly on one hologram and partly on a different hologram, it can only be retrieved by having both holograms. If these are kept separately, the chance of unauthorized snooping is reduced. And additional security can be built in by using different optical arrangements to record each hologram. This is an interesting possibility, as there is great concern about the problems of protecting personal records held on computers.

Another area in which holography offers great promise is the testing of precision engineering components. A standard component can be compared extremely accurately with a supposedly identical one. This is done by superimposing holographic images of the standard component and the component under test. Where the two components differ the interference patterns on their holograms are different. This shows up as a set of light and dark bands (known as *interference fringes*) across the superimposed images. Each fringe represents a difference of one wavelength of light (for example, half a micrometre) between the standard component and the copy. The advantage of this technique is that objects can be compared in three dimensions and thus engineers can use it for controlling the quality of complicated components.

Another, similar, use of holography in mechanical engineering is in the study of how components perform in action. This can be done by comparing the hologram of a component at rest with a hologram taken when the component is working. This method allows engineers to make detailed studies of the performance of machine parts and hence to improve their design. It is also possible to study models of complicated constructions, such as bridges and aircraft wings, to get a better understanding of how they respond to stresses and strains.

If the advantages of holography are combined with the ability of lasers to produce very short pulses, it is possible to produce a very powerful form of high speed photography. Holograms produced in this way can freeze high speed motion and allow scientists to study, in three dimensions, the movement of streams of particles and fluids. This technique has been used to study such things as the airflow over aircraft wings and the flow of petrol/air mixtures in engine carburettors.

Holograms can also be used to measure the tiny changes in size that occur very slowly in some objects. A hologram records the position of all the points on an object with the precision of a fraction of a wavelength of the laser light used. So later holograms of the same object can be used to measure minute changes in size. When two images of the object taken at different times are superimposed, the interference pattern shows how much the object has changed. This technique has been used to study such things as the growth of crystals and the deterioration of old paintings.

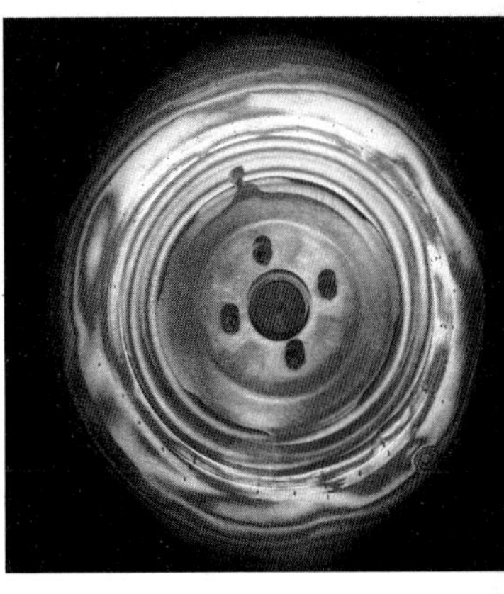

Above: An image of a tyre made by superimposing a holographic image of the unpressurized tyre on a holographic image of the tyre under pressure. The differences between the images show up as interference fringes (light and dark bands) that marks the distortions in the pressurized tyre. A bad distortion can be seen in the 5 o'clock position.

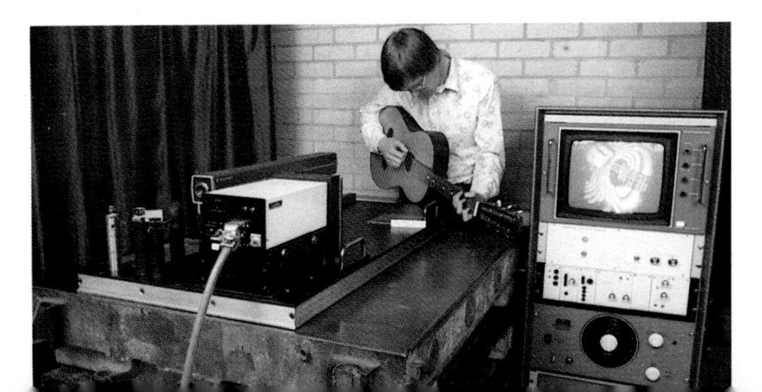

It is possible to create television pictures that show the distortion in moving objects. A hologram of the object is made in the usual way. But instead of being recorded on a photographic plate, it is picked up by a television camera. The hologram is then compared electronically with a previously recorded hologram of the object at rest. Differences between the two holograms show up as contour lines on the television screen. Here, the apparatus is recording the distortion produced in a vibrating guitar.

Glossary

Absorption spectrum is a SPECTRUM in which certain wavelengths have been absorbed by a substance through which the light has passed. These wavelengths show up as dark lines (absorption lines) in the spectrum or as troughs on a graph. They are the wavelengths that are absorbed as the atoms or molecules of the substance change from a low energy level to a higher energy level.

Amplitude of an ELECTROMAGNETIC RADIATION is the total height of the waves. The greater the amplitude, the greater the intensity of the radiation; in the case of light, brightness increases with amplitude.

Atom is the smallest particle of an element that can take part in a chemical reaction. An atom consists of ELECTRONS, NEUTRONS and PROTONS.

Coherent light is light in which all the waves are in phase; that is their peaks and troughs coincide and reinforce each other.

Electromagnetic radiation is energy in the form of waves that consist of an electric field at right angles to a magnetic field. Electromagnetic waves include a whole range of different wavelengths that make up the electromagnetic spectrum (see page 12).

Electron is the smallest particle in an atom. Electrons have almost no mass and are negatively charged. Every electron normally orbits the nucleus (PROTONS and NEUTRONS) of the atom at a particular distance.

Electronic energy levels result from the different electron orbits in an atom. An atom has a "ground state" energy level when its electrons are in their normal orbits. But if the atom is subjected to electromagnetic radiation of particular wavelengths its electrons absorb energy and are boosted to higher orbits, resulting in the atom having a higher energy level.

Emission spectrum is a SPECTRUM that shows the wavelengths given out by a source of light. These wavelengths show up as bright lines in the spectrum (or peaks on a graph) and are the wavelengths emitted as the atoms or molecules of the substance change from a high energy level to a lower energy level.

Energy levels See ELECTRONIC ENERGY LEVELS, ROTATIONAL ENERGY LEVELS.

Frequency of an electromagnetic radiation is the number of waves that pass a given point in one second.

Harmonic generation is the multiplying up of the frequency of laser light. By passing laser light through a special material it is possible to produce harmonics of the original frequency; that is, the laser light that emerges from the material has exactly twice or three times the original frequency.

Inertial guidance system measures all the changes in movement of a ship or aircraft. Three gyroscopes, at right angles to each other, detect when the craft speeds up, slows down or changes direction. This information is fed to a computer, which keeps an accurate record of the craft's position.

Interference occurs when the waves of two beams of light combine and tend either to reinforce each other or to cancel each other our.

Interferometer is an instrument that divides a beam of light into two or more beams and recombines them to produce INTERFERENCE.

Isotopes are two or more forms of the same element. The atoms of different isotopes contain the same number of protons but different numbers of neutrons. Hence their masses are different.

Molecule is the smallest particle of a substance that is capable of existing independently. A molecule may consist of one, two or more atoms of the same element or two or more atoms of different elements.

Neutron is one of the two kinds of particle (see also PROTON) found in the nuclei of all atoms except those of hydrogen. A neutron has no electric charge but has an appreciable mass.

Photon A small "packet" of electromagnetic radiation. It can be visualized as a short train of waves of a particular wavelength.

Proton is one of the two kinds of particle (see also NEUTRON) found in the nuclei of all atoms. A proton has a positive electric charge equal and opposite to the charge of an electron. A proton has an appreciable mass, which is very slightly less than that of a neutron.

Radioisotope A radioactive isotope; an isotope whose atoms decay by emitting particles to become atoms of another element.

Rotational energy levels of a molecule are due to the rotation of two atoms around each other. This rotation has a normal (fundamental) frequency. But if the molecule is excited by energy from an outside source, the atoms may rotate at one of the harmonics of the fundamental frequency—two, three or more times the normal frequency. The molecule therefore has several possible rotational energy levels.

Semiconductor is a material that has properties that lie between those of conductors and insulators. Unlike a metallic conductor, the electrical conductivity of a semiconductor is normally quite low. But its conductivity can be increased by exposing it to heat or light. Semiconductors are usually crystals of such substances as germanium, selenium, silicon, indium and gallium arsenide.

Spectrum The result of splitting electromagnetic radiation, such as light, into the wavelengths of which it is made up. The spectrum of white light is a range of colours and each part of the spectrum is a different wavelength.

Wavelength of an electromagnetic radiation is the distance between successive crests (or troughs) of the train of waves.

Index

Page numbers in *italics* indicate an illustration.

ACKNOWLEDGEMENTS
Cover London Planetarium. Title Bell Laboratories. Pages 6 & 8 Photri. 10 The Who Group, Shepperton Studios. 16 Hughes Aircraft Co. 17 top left Bell Laboratories, top right & bottom Hughes Aircraft Co. 18 Coherent UK Ltd. 19 Central Electricity Research Laboratories. 20 left UKAEA Culham Laboratory, right AERE Harwell. 22 top Office of Earthquake Studies, USA, Bottom Nasa. 23 Nasa. 24 Lasergage Ltd. 25 Mowlem & Co. 26 National Physical Laboratory. 27 & 28 Sperry. 32 Photri. 37 John Brophy, University of Birmingham. 38 Hughes Aircraft Co. 39 & 40 top right Lasercut Products Ltd. 40 top left & bottom UKAEA Culham Laboratories. 41 Coherent UK Ltd. 44 top left & bottom Department of Energy, Washington, top right Rutherford & Appleton Laboratories. 45 Department of Energy, Washington. 46 Hughes Aircraft Co. 47 Bell Laboratories. 49 Sperry. 50 top Photri, bottom Bell Laboratories. 51, 52 & 53 left British Telecom. 53 right Bicc Ltd. 54 Bell Laboratories. 55 top IBM Ltd, bottom Paul Brierley. 56 Hughes Aircraft Co. 57 The Who Group, Shepperton. 58 top Paul Brierley, bottom Photri. 59 University of Loughborough. Back cover Coherent UK Ltd.

Picture research Penny Warn

The Battle of Downing Street

In memory of Charlotte

THE BATTLE OF DOWNING STREET

PETER JENKINS
Political Columnist of The Guardian

CHARLES KNIGHT & CO LTD
LONDON • 1970

CHARLES KNIGHT & CO LTD
11-12 BURY STREET, LONDON, E.C.3.

Printed in Great Britain by
BKT City Print Ltd,
A member of the Brown Knight & Truscott Group

SBN 85314 068 5 (hard cover)
SBN 85314 069 3 (soft cover)